Preface

It's autumn 2016. As this book is getting ready to go to press, news sources (both computing and mainstream) start reporting a massive distributed denial of service attack against Dyn's internet infrastructure—launched from compromised devices that make up the Internet of Things (IoT). A security colleague jokingly asks me whether I arranged the attack as a promotion for this book. A humanities colleague asks me why there are computers inside baby monitors. A science reporter asks what I think—and sadly, I fear this is just the beginning. In autumn 2016, the smart things were only attacking traditional Internet of Computers (IoC) things, with consequences confined to computing. We may have a future where IoT and IoC things alike attack the IoT, with consequences in whatever real-world infrastructure the IoT things augment.

The Internet of Risky Things grew out of a "Sophomore Summer" course I developed and taught at Dartmouth and draws on my lifetime of experience in the interaction of security and information technology and society (including healthcare, investment banking, and 10 years working on cybersecurity in the smart grid—and even e-government at the dawn of the web). This book implicitly assumes a general college education, familiarity with the internet as a user, and the context of citizenship in the world; however, I also see the book being of value to audiences with specialties in computer science, management, research, and education.

The coming IoT distributes computational devices massively in almost any axis imaginable and connects them intimately to previously noncyber aspects of human life. Analysts predict that by 2020 we will have over 25 billion networked devices embedded throughout our homes, clothing, factories, cities, vehicles, buildings, and bodies. As a result, everyone in the developed world will be using the IoT. Anyone who drives or rides in a car will be sharing the road with self-driving cars and depending on smart infrastructure governing the control of traffic lights and the storage and movement of gasoline. Anyone who connects something to the electric grid will be using the IoT. Anyone in a hospital, or buying a new appliance, or in a new commercial house or building will be using it.

In a nutshell: any thinking citizen who plans to still be around in 10 years should be worrying about these issues now.

O'Reilly Safari

Safari (formerly Safari Books Online) is a membership-based training and reference platform for enterprise, government, educators, and individuals.

Members have access to thousands of books, training videos, Learning Paths, interactive tutorials, and curated playlists from over 250 publishers, including O'Reilly Media, Harvard Business Review, Prentice Hall Professional, Addison-Wesley Professional, Microsoft Press, Sams, Que, Peachpit Press, Adobe, Focal Press, Cisco Press, John Wiley & Sons, Syngress, Morgan Kaufmann, IBM Redbooks, Packt, Adobe Press, FT Press, Apress, Manning, New Riders, McGraw-Hill, Jones & Bartlett, and Course Technology, among others.

For more information, please visit *http://oreilly.com/safari*.

How to Contact Us

Please address comments and questions concerning this book to the publisher:

O'Reilly Media, Inc.
1005 Gravenstein Highway North
Sebastopol, CA 95472
800-998-9938 (in the United States or Canada)
707-829-0515 (international or local)
707-829-0104 (fax)

We have a web page for this book, where we list errata, examples, and any additional information. You can access this page at *http://shop.oreilly.com/product/0636920052784.do*.

To comment or ask technical questions about this book, send email to *bookquestions@oreilly.com*.

For more information about our books, courses, conferences, and news, see our website at *http://www.oreilly.com*.

Find us on Facebook: *http://facebook.com/oreilly*

Follow us on Twitter: *http://twitter.com/oreillymedia*

The Internet of Risky Things

Trusting the Devices That Surround Us

Kate —
Thanks for telling us about Kansas City!
— Sean

Sean Smith

Beijing · Boston · Farnham · Sebastopol · Tokyo O'REILLY®

The Internet of Risky Things

by Sean Smith

Printed in the United States of America.

Published by O'Reilly Media, Inc., 1005 Gravenstein Highway North, Sebastopol, CA 95472.

O'Reilly books may be purchased for educational, business, or sales promotional use. Online editions are also available for most titles (*http://safaribooksonline.com*). For more information, contact our corporate/institutional sales department: 800-998-9938 or *corporate@oreilly.com*.

Editors: Brian Jepson and Jeff Bleiel
Production Editor: Shiny Kalapurakkel
Copyeditor: Rachel Head
Proofreader: Amanda Kersey
Indexer: Judy McConville
Interior Designer: David Futato
Cover Designer: Karen Montgomery
Illustrator: Rebecca Demarest

February 2017: First Edition

Revision History for the First Edition

2017-01-13: First Release

See *http://oreilly.com/catalog/errata.csp?isbn=9781491963623* for release details.

978-1-491-96362-3

[LSI]

Contents

Watch us on YouTube: *http://www.youtube.com/oreillymedia*

Acknowledgments

I would like to thank Sergey Bratus, John Erickson, Ross Koppel, Bill Nisen, and the students in the Dartmouth Trust Lab for their helpful discussion, and my colleagues and partners in the DOE/NSF/HHS *Trustworthy Cyber Infrastructure for the Power Grid* and *Cyber Resilient Energy Delivery Consortium* projects for giving me a chance to learn much about and help a little with one of the world's largest cyber-physical systems.

Thanks also to Jamie Coughlin, Katharina Daub, and Michael Wooten for their help in setting up the "Risks of the Internet of Things" living/learning course—and to the students in the course for making it all happen. And to Gretchen, for her patience and support.

Brave New Internet

The coming *Internet of Things* (IoT) distributes computational devices massively in almost any axis imaginable and connects them intimately to previously non-cyber aspects of human life.[1] Analysts predict that by 2020 humanity will have over 25 billion networked devices, embedded throughout our homes, clothing, factories, cities, vehicles, buildings, and bodies. If we build this new internet the way we built the current *Internet of Computers* (IoC), we are heading for trouble: humans cannot effectively reason about security when devices become too long-lived, too cheap, too tightly tied to physical life, too invisible, and too many.

This book explores the deeper issues and principles behind the risks to security, privacy, and society. What we do in the next few years is critical.

Worst-Case Scenarios: Cyber Love Canal

Many visionaries, researchers, and commercial actors herald the coming of the Internet of Things. Computers will no longer look like computers but rather like thermostats, household appliances, lightbulbs, clothing, and automobiles; these embedded systems will permeate our living environments and converse with one another and all other networked computers.

What's more, these systems will intimately interact with the physical world: with our homes, schools, businesses, and bodies—in fact, that's the point. In the visions put forth, the myriad embedded devices magically enhance our living

1 Indeed, strictly speaking, it's not correct to say the IoT is *coming*; as examples throughout this book show, it's already here. What's coming is much, much more of it—and as one colleague observes, "In making predictions, we always tend to overestimate short-term impact and underestimate long-term impact."

environments, adjusting lights, temperature, music, medication, fuel flow, traffic lighting, and elevators.

It's tempting to insert a utopian sketch of a "typical day in 2025" here, but there are plenty of sketches out there already. It's also tempting to insert a dystopian sketch of what might happen if a malicious adversary mounts a devious attack, but that's been done nicely too—see Reeves Wiedeman's article "The day cars drove themselves into walls." It's a scenario that could happen based on what already has [111].

However, I will offer an alternative dystopian vision of this future world. Every object in the home—and every part of the home—is inhabited by essentially invisible computational boxes that can act on the physical environment. But rather than being helpful, these devices are evil, acting in bizarre, dangerous, or unexpected ways, either chaotically or coordinated in exactly the wrong way. We can't simply turn these devices off because no one knows where the off switches are. What's worse is that we don't even know where the devices are!

This vision of dark magic might inspire us to look to horror novels or science fiction for metaphors. However, real life has given us a better metaphor: environmental contamination. We've seen buildings contaminated by lead paint and asbestos and rendered unusable by chemical spills at nearby dry cleaners, a research lab rendered uninhabitable by toxic mold, and Superfund sites called *brownfields* that can't be built on (at least, without often expensive remediation). In all these cases, technology (usually chemical) intended to make life better somehow backfired and turned suburban utopias into wastelands.

What happened at the Pearl Harbor naval base is widely known. However, Americans over 50 might also remember Love Canal, a neighborhood of Niagara Falls, New York, that became synonymous with chemical contamination catastrophe. Vast amounts of chemical waste buried under land that later supported homes and schools led to massive health problems and the eventual evacuation and abandonment of most of the neighborhood.

What some might term a "cyber Pearl Harbor"—a coordinated, large-scale attack on our computational infrastructure—would indeed be a bad thing.[2] However, we should also be worried about a "cyber Love Canal"—buildings and neighborhoods, not to mention segments of our cyberinfrastructure, rendered

2 However, despite the impact of the Pearl Harbor attack, one should keep in mind that it was just an attack on one base, and only on some of the ships at that base; a modern cyber version would likely not have such a limited scope.

uninhabitable by widespread "infection" and loss of control of the IoT embedded therein. The way we build and deploy devices today won't work at the scale of the envisioned IoT and will backfire, like so many hidden chemical dumps. Continuing down this path will similarly lead to "cyber brownfields."

"Things don't work" is the security idealist's standard rant. A more honest observation is that, although flawed, things work well enough to keep it all going. For the most part, we know where the machines are—workstations and laptops in offices, servers in data centers. The operating system (OS) and application software are new enough to still be updated, and machines are usually expensive enough to justify users' attention to maintenance and patching before too many compromises happen.[3]

The current IT infrastructure is compromisable and compromised, with occasional lost productivity and higher fraud losses amortized over a large population—yet life goes on, mostly. The fact that I am writing these words on networked machines while the web continues to work proves that.

However, in the IoT, the numbers, distribution, embeddedness, and invisibility of devices will change the game. Suppose we build the IoT the same way we built the current internet. When the inevitable input validation bug is discovered, there will be orders of magnitude more vulnerable machines. Will embedded machines be patchable? Will anyone think to maintain inexpensive parts of the physical infrastructure? Will machines and software last longer than the IoT startups that create them? Will anyone even remember where the machines are? Imagine a world in which we needed to update each door, each electrical outlet, and perhaps even each lightbulb on Patch Tuesday.

When the inevitable happens, what will a compromised machine in the IoT be able to do? It's no longer just containing data; it's controlling boiler temperatures, elevator movement, automobile speed, fish tank filters, and insulin pumps. Consider the effects of denial of service on our physical infrastructure. In winter where I live, temperatures of –15°F are common. How many burst pipes and damaged buildings—and maybe even deaths—would we have had if a virus shut down all the heating systems? The recent nonfiction book *Five Days at Memorial* by Sheri Fink chronicled the horrors of being trapped in a New Orleans hospital when Hurricane Katrina shut down basic infrastructure, including electricity,

3 At least, one likes to think so; friends and colleagues point out examples of mission-critical computing running on obsolete operating systems and/or with source code no longer available.

transport, and communication [31]. Could an infection in the IoT cause similar infrastructure loss?

Fail-stop is bad enough, but compromised machines in the IoT can do more than simply stop; they can behave arbitrarily. What havoc might happen when elevators, automobiles, and door locks start behaving unpredictably? A decade ago, a compromise at my university led to a large server being coopted to distribute illegal content—annoying, but relatively harmless. What could happen when schools, homes, apartment buildings, and shopping malls are full of invisible, forgotten, and compromised smart devices?

What's Different?

"We already have an internet," some readers may wonder, "and it works pretty well. Why the fuss about a new one?" This chapter addresses that question. What's different about the IoT?

Kevin Ashton, credited with inventing the term "IoT," sees the connection to humanity as the distinguishing factor [5]:

> *The fact that I was probably the first person to say "Internet of Things" doesn't give me any right to control how others use the phrase. But what I meant, and still mean, is this: Today computers—and, therefore, the Internet—are almost wholly dependent on human beings for information. Nearly all of the roughly 50 petabytes (a petabyte is 1,024 terabytes) of data available on the Internet were first captured and created by human beings—by typing, pressing a record button, taking a digital picture or scanning a bar code. Conventional diagrams of the Internet include servers and routers and so on, but they leave out the most numerous and important routers of all: people.*

Ahmed Banafa goes further—the next stage in the evolution of computing systems, the IoT connects to *everything* [6]:

> *The Internet of Things (IoT) represents a remarkable transformation of the way in which our world will soon interact. Much like the World Wide Web connected computers to networks, and the next evolution connected people to the Internet and other people, the IoT looks poised to interconnect devices, people, environments, virtual objects and machines in ways that only science fiction writers could have imagined.*

This latter definition is the sense in which I use the term in this book. In the IoT, we are layering computation on top of everything—and then interconnecting it.

Because it thus scales up from the IoC in so many dimensions, the IoT becomes something new. Even in the IoC, internet pioneer Vint Cerf saw the newness created by scale [50 p. 29]:

> *...It's difficult to envision what happens when extremely large numbers of people gain access to a technology, such as the Internet. It's a bit like being the inventor of the automobile and imagining a few dozen of them, not knowing how 50 million or 100 million of them would affect the attitudes, customs, behavior and actions of the entire country...and the world.*

However, the IoT will be many orders of magnitude larger than even the current internet and interconnect devices far smaller. What's more, the amount of *data* these devices generate will also be orders of magnitude greater; thanks to its embedded machines, a Boeing 737 flying from Los Angeles to Boston generates more bytes of data than the Library of Congress collects in one month.

LIFETIMES

Another dimension of scale is time. The devices in the IoT will persist far longer than the current thinking allows for. Unreachable or forgotten devices will disrupt the "penetrate and patch" model governing the current internet, and devices may have baked-in cryptography persisting for decades beyond its security lifetime.

The devices may even outlive the companies and enterprises responsible for maintaining them. Consider some data points:

- The film *2001: A Space Odyssey*, whose creators in 1968 put much thought into imagining the world of 2001, is noted for its "curse": businesses whose logos are featured prominently in the film ended up disappearing in the real world. Will our predictions today be any better? Who will be around in 30 years to issue software updates to the smart appliances I buy today?
- In April 2016, *Business Insider* reported how Google Nest "is dropping support for a line of products—and will make customers' existing devices

completely useless" [86]. In the IoT, how will the often short lifetimes of IT systems mesh with the typically long lifetimes of physical ones?

THE IOT IN THE PHYSICAL WORLD

The IoT will intimately tie the cyberworld to physical reality, to a degree never before achieved, and all of these new dimensions will disrupt and subvert the way technologists, policy makers, and the general public reason about the IT infrastructure. The remainder of this chapter explores these connections and implications:

- Why software requires continual maintenance, and what this means when we change from the IoC to the IoT.
- The impact that the IoT, so much larger in scale, may have on the physical world.
- The impact of the physical world on the IoT (the connection works both ways).
- Worst-case scenarios of an IoT-enabled *cyber Pearl Harbor* (or *cyber Love Canal*).
- How we, as a society, might move forward safely.

Inevitable and Unfortunate Decay

When it comes to the physical world, *engineering* generally works—with software, not so much. A friend in grad school once quipped, "Software engineering is the field of the future, and it will always be the field of the future." Although this was several decades ago, it is unfortunately still true.

When it comes to building physical things out of materials such as wood, steel, and concrete, humanity has figured out how to select the right materials and then assemble them into things like houses and bridges, with a fairly good estimate of how long these houses and bridges will hold up. It does not happen that every few weeks, an announcement goes out that all houses built with a certain kind of door suddenly need to have the doors replaced; it is not the general custom for people to avoid going into houses more than 20 years old because they are worried they will suddenly collapse or be filled with dangerous criminals, simply because the owners neglected to repaint them every "paint Tuesday."

However, when it comes to things humans build in software, these scenarios —ridiculous for physical infrastructure—are standard practice. It is extremely hard to build software systems that, with high confidence, will last a few years (let alone decades) without the discovery of bugs and critical security issues. Reasons effective *software engineering* is so hard include:

- The sheer number of "moving parts" (e.g., it takes at least a half dozen Boeing 747 airplanes to have as many physical parts as lines of code in a basic laptop OS).
- The way these parts compose and interact (it is unlikely for a part in one airplane to reach out and subvert the functioning of a different part in a different airplane—but this happens all the time in software modules).
- The economic forces that can make software products be rushed to market with insufficient testing.
- The fact that, thanks to the internet, the attack surface (perimeter) on a piece of software may be exposed to all the world's adversaries, all the time.

Nonetheless, this is the reality. Systems that do not receive regular patching are assumed to be compromised—leading to a *Through the Looking-Glass* world where users and administrators must keep running to stay in the same place (Figure 1-1). What's even stranger about this, from the physical engineering metaphor, is that it's usually not the case that the components suddenly break and must be fixed. Rather, usually, the components were *always broken*; it's only that the defenders (and hopefully the adversaries also) just learned that fact.

Figure 1-1. The need for software updates means we must continually run just to stay in the same place. (John Tenniel illustration for Through the Looking-Glass, public domain.)

Indeed, on the day I was writing this, researchers from cyber security company Check Point announced that Facebook's Chat service and Messenger application have flaws in how participants construct URLs for communication, and that an adversary can use these flaws to take over conversations [13]. Yesterday, the day before, and even back in 2008 when Chat was announced, this service was considered secure—but the zero-day vulnerabilities were always there.

ZERO-DAYS AND FOREVER-DAYS

The IoC has thus come to depend on the *penetrate and patch* paradigm. We keep running to stay in the same place, to reduce the risk of attack via *zero-days*—holes that adversaries know about but defenders do not. In just one random two-week period in 2016, the US Department of Homeland Security (DHS) Industrial Control Systems Cyber Emergency Response Team announced 11 critical zero-days:

- In a wireless networking device used in "Commercial Facilities, Energy, Financial Services, and Transportation Systems" internationally
- In an embedded computer used in "Chemical, Commercial Facilities, Critical Manufacturing, Emergency Services, Energy, Food and Agriculture,

Government Facilities, Water and Wastewater Systems" and other sectors internationally

- In building and automation systems from two vendors used in "Commercial Facilities" internationally
- In power grid components from three vendors, used internationally
- In other industrial control systems networking equipment from four vendors, used in "Chemical, Critical Manufacturing, Communications, Energy, Food and Agriculture, Healthcare and Public Health, Transportation Systems, Water and Wastewater Systems, and other sectors" internationally

Critical infrastructure components are in need of urgent update, just to remain as safe as the community thought they were when first deployed.

The widespread use of common, commodity software components in the IoT (and IoC) raises the specter of what my own lab has called *zero-day blooms*: like algae blooms, rapidly spreading exploitation of newly discovered holes might threaten a wide swath of infrastructure [80]. In just one month in 2016, Ubiquiti Networks announced that a hole in its energy-sector wireless equipment was being exploited in installations worldwide [46], and security firm Trustwave announced that an entity was selling a zero-day vulnerability allegedly "affecting all Windows OS versions"—yes, all of them [20].

Consider the implications if, in 10 years, a similar zero-day surfaces in the commodity OS used by a larger part of the IoT. What will be caught in the zero-day bloom? If the IoT requires "penetrate and patch" to remain secure, how will that mesh with the physical world to which the IoT is tied? Indeed, in the Ubiquiti incident just mentioned, the vulnerability had been discovered and a patch released nearly a year earlier—yet the hole persisted. For another example, content security software company Trend Micro wrote, in December 2015, of 6.1 million smart devices at risk due to vulnerabilities for which patches had existed since 2012 [119]. In January 2016, the *Wall Street Journal* reported on 10 million home routers at risk because they used unpatched components from 2002 [106]. Indeed, the difficulty—or impossibility—of patching in the IoT has caused some analysts to suggest a new term: *forever-days* [45], zero-days that are permanent.

THE FIX IS IN?

Because of this inevitable decay, software systems require continual maintenance.

In the IoC

Even in the IoC, where computers look like computers and usually have human attendees, maintaining software is an ongoing challenge. Consider the recent case of Microsoft trying to upgrade machines to Windows 10. The *Guardian* catalogued some of the complaints [42]:

> *Scores of users have posted...to complain about Windows 10 automatically installing, seemingly without asking, and often in the middle of doing something important.*
>
> *[A] Reddit user...posted a warning...after it..."bricked" his father's computer....*
>
> *"In some cases the upgrade went OK and the user is just really confused. In others, Windows 10 is asking for a login password the user set years ago and hasn't used since, that was fun. In still another it's screwed up access to their shared folders."*

A complicating factor here is that although Microsoft tried to get the user's permission before starting an upgrade, there was confusion about the interface. As the BBC put it [61]:

> *Clicking the cross in the top-right hand corner of the pop-up box now agrees to a scheduled upgrade rather than rejecting it. This has caused confusion as clicking the cross typically closes a pop-up notification.*

Even here in the IoC, maintenance has not meshed well with how IT is tied to the real world. *Engadget* and others noted a Windows 10 upgrade dialog box popping up on the weather map during a television news broadcast [101]. The *Register* wrote that an anti-poaching organization in the Central African Republic had its (donated) laptops begin automatic upgrades, overwhelming its (expensive) satellite network link and rendering the machines inoperable [102].

In the IoC, effective maintenance of software at other granularities than operating systems has also been a problem. The *BIOS*, and other low-level firmware that runs before the OS boots, provides some examples. Softpedia writes that the ASUS motherboards automatically check for and then install updates—

but do so without checking the authenticity of the upgrade [18]. Researchers at Duo found problems with low-level software updating from many other vendors as well [57]. (Chapter 4 will consider further this design pattern of "failure to authenticate.") Managing updates for components with long lifetimes and many generations can also be problematic, as Oracle learned when the Federal Trade Commission (FTC) sanctioned it for its Java update policy, which fixed recent versions but left older, vulnerable versions exposed [15].

Into the IoT

The challenges that made these software maintenance issues in the IoC problematic will be even stronger in the IoT. Even now, in the early stages of the IoT, maintenance problems emerge:

- In 2014, Toyota required Prius owners to bring cars to dealers to fix software bugs that could "cause vehicles to halt while being driven" [32].
- In 2015, Jaguar Land Rover required customers to bring cars to dealers to fix software bugs pertaining to door locks [121].
- In 2015, BMW patched a security flaw via automatic download [76], as did Tesla [108].
- In 2015, a software maintenance problem on an Airbus A400M led to an even worse scenario: a crash killing four [38].
- After the much-publicized *Wired* demonstration of remote takeover of a Jeep [49], Fiat Chrysler pushed out a software update by physically mailing a USB drive to owners [48] and asking them to install it—a mechanism with questionable security and effectiveness. (On the other hand, Fiat Chrysler also "applied network-level security measures on the Sprint cellular network that communicates with its vehicles" [58].)
- In June 2016, patching via automatic download did not go so well for Lexus owners, who found the update rendered their navigation systems and radios inoperable [44].
- In February 2016, a Merge Hemo (a computerized "documentation tool" for cardiac catheterization) crashed during a heart operation because automatic antivirus scanning had started [19].

So far, software maintenance in the IoT has had a mixed record.

As noted earlier, another example of the mismatch between IT and physical reality is relative lifetimes. Things in one's house (or in bridges or cars) may last longer than the supported lifetime of a given IoT component. For one example, Trend Micro recently noted [39]:

> *Our researchers have found a hole in the defenses of the systems on chips (SoCs) produced by Qualcomm Snapdragon that, if exploited, allows root access.... So far, Trend Micro security experts have found this vulnerability on the Nexus 5, 6, 6P and the Samsung Galaxy Note Edge. Considering the fact that these devices no longer receive security updates, this is concerning news for anyone who owns one of these phones. However, smartphones aren't the only problem here. Snapdragon also sells their SoCs to venders producing devices considered part of the IoT, meaning these gadgets are just as at risk.*

Referring to this general problem of IoT patching, FTC Chairwoman Edith Ramirez lamented how "it may be difficult to update the software or apply a patch—or even to get news of a fix to consumers" [73]. In one extreme example, in November 2015, Orly Airport (near Paris) was brought "to a standstill" because of software critical to its operations that was running on Windows 3.1 [74]. This OS is so "prehistoric," according to Longeray, that patches are no longer even available; Alice cannot run fast enough.

The IoT's Impact on the Physical World

One principal way the IoT differs from the IoC is the intimate connection of the IoT to physical reality. This connection amplifies the consequences of IT—expected behavior, unexpected behavior, malicious action.

For an example of troubling consequences for teachers (like me), Amazon sells some interesting smart watches [55]. As the *Independent* reported in March 2016:

> *"This watch is specifically designed for cheating in exams with a special programmed software. It is perfect for covertly viewing exam notes directly on your wrist, by storing text and pictures. It has an emergency button, so when you press it the watch's screen display changes from text to a regular clock, and blocks all other buttons," the seller wrote.*

However, there are already many examples of deeper consequence.

HOUSES

Perhaps the most immediately tangible aspect of "physical reality" are the homes and apartments we live in. The marketplace is already filled with smart televisions, thermostats, light switches, door locks, garage door openers, refrigerators, washing machines, dryers (and probably many other things I've missed).

One way the IoT has been negatively impacting this intimate physical reality is simply by not working. Adam Clark Estes (in the aptly titled "Why is my smart home so fucking dumb?") writes [30]:

> *I unlocked my phone. I found the right home screen. I opened the Wink app. I navigated to the Lights section. I toggled over to the sets of light bulbs that I'd painstakingly grouped and labeled. I tapped "Living Room" —this was it—and the icon went from bright to dark. (Okay, so that was like six taps.)*
>
> *Nothing happened.*
>
> *I tapped "Living Room." The icon—not the lights—went from dark to bright. I tapped "Living Room," and the icon went from bright to dark. The lights seemed brighter than ever.*
>
> *"How many gadget bloggers does it take to turn off a light?" said the friend, smirking. "I thought this was supposed to be a smart home."*
>
> *I threw my phone at him, got up, walked ten feet to the switch. One tap, and the lights were off.*

Terence Eden similarly wrote of "The absolute horror of WiFi light switches" [28].

Adding interconnected smartness to a living space also opens up the potential of malicious manipulation of that space via manipulation of the IT. For example, Matthew Garrett wrote of some penetration testing he did while staying at a hotel that had "decided that light switches are unfashionable and replaced them with a series of Android tablets" [40]. With straightforward Ethernet sniffing, he discovered he could use his own machine to turn on and off lights, turn on and off the television, open and shut curtains—in any room in the hotel.

Numerous competing vendors already offer *hubs* to serve as the central connectivity point for the smart home; others, such as Samsung and SmartThings, are trying to turn one of the smart appliances into that hub (e.g., [24]). These hubs can also provide a vector for external adversaries to enter the domicile. In

2015, researchers from Veracode found many vulnerabilities in such commercial offerings [82]. In 2016, researchers from Vectra Threat Lab found ways to use web-connected home cameras to penetrate the home network [122]. Many years ago, my own lab found such holes with set-top boxes our university deployed in dorm rooms.

In the IoC, security specialists used to have to explain that the reason it's important to lock one's computer even if one doesn't lock one's front door is that one's front door only opens into the neighborhood, while the computer opens to the entire world. Thanks to the IoT, the front door now opens to the entire world as well.

CARS

For many people, after the home, one's car is one's castle: an environment that's personal and part of everyday life. The penetration of IT and networks into the automobile thus provides another vector for the IoT to disrupt physical reality—unpleasantly.

Random failures have caused trouble already. In 2003, Thailand's finance minister was trapped in his BMW when its computer failed [4]. In 2011, a young man died in France when he was unable to exit a locked car, because (by design) the car could not be opened from the inside if it had been locked from the outside [1]. In June 2015, a man and his dog died in Texas when he was unable to open the door or window in his Corvette, due to a battery failure [92]: "Police believe [he] made a valiant effort to escape, and possibly died while looking through the car's manual."

Malfeasance can also happen. In February 2015, surveillance video shows a thief opening a locked Audi and stealing a $15,000 bicycle from inside—apparently by just touching the car [43]. In May 2015, researchers in Germany analyzed BMW's ConnectedDrive wireless transmissions and found vulnerabilities that would let an external adversary open a locked car [99]. In August 2015, Threatpost wrote of researcher Samy Kamkar's tools enabling adversarial subversion of various remote car unlocking features [33]. One wonders: could there be a connection?

These incidents are probably overshadowed by the larger story of researchers breaking into the internal networking of cars. Here are some examples:

- In 2010, researchers from the University of South Carolina and Rutgers identified security vulnerabilities in how tire pressure monitors communicate wirelessly with modern cars [90].
- In 2010 and 2011, researchers from UC San Diego and the University of Washington identified—and demonstrated on real cars under controlled conditions on closed roads—many ways adversaries, both remote and local, can cause havoc [65, 14].
- In February 2015, a teenager built a wireless device that could "remotely control...headlights, window wipers, and the horn" and "unlock the car and engage the vehicle's remote start feature" [74]. Interestingly, these features were deemed "non-safety related."

Nonetheless, it wasn't until July 2015 [49], when two researchers remotely took control of *Wired* reporter Andy Greenberg's Jeep—live, on the highway—that the potential of adversarial exploitation of the IoT's penetration into automobiles caught the public's attention. In the *New York Times* opinion pages the next month, Zeynep Tufekci quipped [105]:

> *In announcing the software fix, the company said that no defect had been found. If two guys sitting on their couch turning off a speeding car's engine from miles away doesn't qualify, I'm not sure what counts as a defect in Chrysler's world.*

Subsequent work has penetrated cars via digital radio [107], text messaging [41], and WiFi [123]. Technicians at the National Highway Traffic Safety Administration (NHTSA) demonstrated "ways of tampering remotely with door locks, seat-belt tensioners, instrument panels, brakes, steering mechanisms and engines—all while the test cars were being driven" [26]. Richard Doherty of The Envisioneering Group noted "there's nothing secure and now we're putting chips into things that go 100 miles per hour and in Germany faster" [11]. Craig Smith, author of *The Car Hacker's Handbook* (No Starch Press) [95], observed the potential of an infected car to then compromise the IoT in the car owner's smart home.

TRAFFIC

Connecting IT to cars in the aggregate can impact how cars, in the aggregate, impact the physical world.

For one example, consider what happened to previously quiet neighborhoods in Los Angeles in late 2014 [89]. Waze, an app using crowdsourced traffic data to help its users find faster routes through town, started routing massive amounts of traffic through these neighborhoods. According to locals, the small residential streets paralleling the busy Interstate 405 freeway became "filled each weekday morning with a parade of exhaust-belching, driveway-blocking, bumper-to-bumper cars." The app caused controversy again in March 2016, when Israeli soldiers using Waze inadvertently entered Palestinian territory, leading to a firefight with fatalities [120].

Other new car technologies may have unclear impacts. Will self-driving cars reduce traffic and accidents, as Peter Wayner [110] and Baidu [62] claim? Or will, as AP reporter Joan Lowy notes [75], the decreased cost and increased ease of driving cause traffic to skyrocket? (Summer 2016's news of a fatality from a Tesla on autopilot adds more wrinkles to the discussion [71].) It would seem as if changing from gasoline-powered cars to electric cars would decrease pollution, but researchers from Dartmouth and elsewhere have shown the picture is much more complicated and depends on the local power infrastructure [54]. Similarly, changing to more efficient self-driving cars might reduce the impact on the environment by reducing the time spent driving solo and driving searching for parking—but researchers from the University of Leeds, the University of Washington, and Oak Ridge National Laboratory have similarly found the picture is more complicated [88].

AIRPLANES

Airplanes are another high-impact transportation medium affected by the IoT.

The planes themselves are distributed cyber-physical systems: computers and networks and sensors and actuators packaged and distributed in a thin metal frame flying through the air and carrying people. Back in the 1980s at Carnegie Mellon University's Software Engineering Institute, I was told about a fighter jet that had a bug in its software—and it was easier to fix the jet to match the software than the other way around. (Unfortunately, I have never been able to find independent documentation of this problem.) In 2011, the *Christian Science Monitor* reported that Iranians had captured a US drone by remotely compromising its navigation system and convincing it to land in Iran [85], bringing to mind

World War II stories of UK attacks on the more primitive navigation systems used by German bombers [56].

In 2007, associate professor in the department of Electrical and Computer Engineering at Carnegie Melon University, Phil Koopman [64] lamented how, as the airplane's computer systems reached to the passenger seats, "Passenger laptops are 3 Firewalls away from flight controls!" In early 2015, security researcher Chris Roberts made headlines by being detained by the FBI for allegedly breaking into and manipulating an airplane's controls from a passenger seat [118]. (Some doubt exists as to how successful he really was.) According to reports, to do this, he first had to physically break into the electronic box under this seat to access the network ports. (Chapter 3 will consider some of the controversy surrounding the concept of *fly-by-wire*.) As Figure 1-2 shows, later in 2015, I took a flight on a commercial 747 that made this physical break-in step unnecessary. (No, I did not attempt an attack.)

Figure 1-2. On this commercial flight, I would not need to break into a physical box to connect to the Ethernet. (Photo by author.)

The penetration of IT into the dependent infrastructure can also have consequences. In April 2015, American Airlines flights were delayed when "pilots' iPads—which the airline uses to distribute flight plans and other information to the crew—abruptly crashed" (apparently due to a faulty app) [81]. In June 2015, United Airlines grounded all of its US flights due to an unspecified problem with "dispatching information" [117]. In June 2015, LOT Polish Airlines needed to ground flights due to an unspecified cyberattack on its ground systems [22]. In August 2015, problems with the air traffic control system in the eastern US significantly disrupted flights to and from Washington and New York [98].

The IoT reaches far.

INFRASTRUCTURE

The penetration of IT into other aspects of the surrounding infrastructure can also have a negative impact.

For example, consider commuting. In 2015, a commuter train in Boston somehow left its station with passengers but with no driver [91]. Also in 2015, Dutch newspapers reported a number of accidents (including fatalities) from, due to some system bug, the lights at road and train crossings showing "green" even though a train was present [37]. In 2015, the *New York Times* reported on the fragile state of IoT-connected traffic lights [84]:

> *Mr. Cerrudo, an Argentine security researcher at IOActive Labs, an Internet security company, found he could turn red lights green and green lights red. He could have gridlocked the whole town with the touch of a few keys, or turned a busy thoroughfare into a fast-paced highway. He could have paralyzed emergency responders, or shut down all roads to the Capitol.*

For other examples, consider industrial control systems. In 2000, a disgruntled technician in Australia hacked the control system of a sewage treatment facility to discharge raw sewage [97]. In 1999, a complex system failure (including an unexplained computer outage) caused a gasoline pipeline in Washington State to discharge gasoline, leading to a deadly fire [94]. In 2013, after finding many industrial control systems exposed on the internet, researchers at Trend Micro built *honeypots* (basically, traps in that they are systems controlled by researchers that appear to be exposed internet-facing industrial control systems) and recorded over 33,000 automated attacks [113, 112]. In January 2015, Rapid7 surveyed the internet and found over 5,000 exposed automated tank gauges at gas stations in the US [52]. In February 2015, Trend Micro found an online gasoline pump that had been renamed "WE_ARE_LEGION," presumably as a nod to Anonymous [3]. Trend Micro also set up gasoline-specific honeypots and recorded more attacks [114].

In 2014, the *New York Times* reported on dozens of cases where factory robots had killed workers, cataloging many cases in gruesome detail [77]. (I would have reproduced them here, but the licensing fees were too expensive.) In 2015, robots killed again, at a Volkswagen factory in Germany [8].

MEDICINE

The IoT also penetrates medical infrastructure.

Security risks from software used in implantable medical devices—such as insulin pumps and pacemaker-style cardiac devices—have been in the public eye for a while, thanks in part to the work of researchers such as Kevin Fu (e.g., [10]). Indeed, fear of such an attack led to Vice President Dick Cheney disabling wireless access to his heart device [63].

However, the devices outside bodies but inside hospitals are also at risk. In 2013, a team of penetration specialists including Billy Rios conducted an engagement at the Mayo Clinic [87]:

> *For a full week, the group spent their days looking for backdoors into magnetic resonance imaging scanners, ultrasound equipment, ventilators, electroconvulsive therapy machines, and dozens of other contraptions.*
>
> *"Every day, it was like every device on the menu got crushed."*

Many similar engagements (e.g., [116, 36]) have had similar findings. In June 2015, researchers from TrapX found that in hospitals, the computers disguised as medical devices "all but invisible to security monitoring systems" were rife with malware, some generic but some apparently designed to exfiltrate sensitive medical data [83]. In 2016, researchers found numerous software holes in the Pyxis drug-dispensing cabinets used in hospitals [124]—interesting in part because malicious misuse of the Pyxis interface had already been used to obtain drugs used to murder patients [47].

(Chapter 8 will revisit challenges in using public policy to address these problems.)

The Physical World's Impact on the IoT

Distributing IT throughout physical reality not only permits the IoT to impact the physical world—it also permits the physical world to impact the IoT.

MISSING THINGS

One avenue to consider is whether or not physical environments provide the connectivity an IoT system expects (often implicitly). Here are a few examples:

- Here at Dartmouth, we had an interesting incident where a seminar speaker drove—with his family—the 125 miles from Boston to Hanover in a rental car that required "phoning home" before the engine would start. When the speaker stopped for a picnic at a scenic location in rural New Hampshire, he was not able to start the car again, because our part of the country is riddled with "dead zones" with no cellular coverage. The only way to solve the problem was to have the car towed to someplace with cell coverage.
- In the spring of 2015, the National Institute of Standards and Technology (NIST) released a report raising concerns about IoT applications that "frequently will depend on precision timing in computers and networks, which were designed to operate optimally without it." The report noted that additionally, "many IoT systems will require precision synchronization across networks" [9].
- In May 2015, the Fairfax, Virginia, public schools had their standardized testing disrupted for 90 minutes due to a problem in internet connectivity [29]. "Some students had to wait in the test environment after they completed their tests until connectivity was restored and they were able to submit the tests."
- At a World Cup ski race in Italy in December 2015, a drone crashed and nearly hit a skier—apparently because interference obstructed the radio channel over which it was controlled [104].
- In May 2016, Terminal 7 at Kennedy International Airport experienced "travel chaos" because of a network outage; check-in lines stretched to over 1,500 people, and counter personnel needed to issue boarding passes by hand [100].

LARGE ATTACK SURFACE

Another avenue to consider is the degree to which the physical exposure of the IoT provides avenues for adversaries to manipulate the IoT by manipulating remote computation nodes. One basic approach is injecting fake data at sensors:

- The current electrical grid (and the emerging *smart grid*) uses distributed sensing of electrical state to balance generation, transmission, and consumption, and to keep things from melting down and blowing up. An area of ongoing concern is whether an adversary can cause the grid to do bad things simply by forging some of this data. At the moment, this issue is in churn: researchers seem to alternate between discovering successful attack strategies and then discovering countermeasures, in part using the fact that the set of all measurements must be consistent with the laws of physics as a "secret weapon" to recognize some kinds of forgery—if a sensor reports something that is not physically possible, then we might conclude that it is lying. (See NIST's 2012 *Cybersecurity in Cyber-Physical Systems Workshop* for a snapshot of this work; Chapter 3 will discuss the smart grid further.)
- When (as discussed earlier) residents of previously quiet Los Angeles neighborhoods were peeved by Waze routing commuter traffic there, they tried to sabotage the app by reporting fake accidents and such; unfortunately for them, massive crowdsourcing enabled Waze to filter out these reports as spurious. (Researchers are making progress on both attacks and defenses here [109].)
- Some companies are starting to offer insurance discounts to employees who exercise regularly, as documented by wearable Fitbit devices. In response, some enterprising engineers at one of these companies have come up with techniques to forge this documentation [59].

The basic surroundings of IT devices may provide surprising attack vectors. In 2015, researchers demonstrated Portable Instrument for Trace Acquisition (PITA) devices that could spy on a nearby computer via the electromagnetic radiation it emits and that can also fit inside a piece of pita bread [60]. More recently, researchers have demonstrated how adversaries over 100 meters away can eavesdrop on and alter communications between a computer and a wireless keyboard or mouse connected to it, thanks to MouseJack vulnerabilities (*https://www.mouse jack.com/*).

JUMPING ACROSS BOUNDARIES

In addition to the physical world influencing the IT infrastructure in surprising ways, one sometimes sees one segment of IT infrastructure influencing another segment—surprising because these segments were previously thought of as independent. The coming IoT is manifesting more of this:

- For some kinds of electric vehicles, the existing power infrastructure in many local neighborhoods in the US cannot support more than one vehicle charging at a time [51]—introducing another argument for a smart grid that can coordinate between vehicles.
- A regional electric utility considering deploying wireless smart meters analyzed the potential impact if a hacker penetrated these meters and decided the benefits outweighed the risks (this was relayed to me in personal communication). However, the analysis considered merely the risk to the power grid IT, and neglected to consider that a hacker inside a meter might redirect its software defined radio (SDR) to subvert other infastructures, such as the cellular network.
- The 2013 compromise of Target credit card data stemmed from credential theft from an HVAC vendor [68]—leading one to wonder why an HVAC vendor needs to connect to the same networks that house the credit card crown jewels.
- January 2014 brought reports of infected smart refrigerators sending over 750,000 malware-laden emails [2].
- In January 2015, Krebs reported, "The online attack service launched late last year by the same criminals who knocked Sony and Microsoft's gaming networks offline over the holidays is powered mostly by thousands of hacked home Internet routers" [66].
- In May 2015, *ITworld* reported how employee web browsing and email use was providing a vector for malware to infect point-of-sale (POS) terminals [25].
- In December 2015, the *International Business Times* hypothesized that a recent internet-wide distributed denial of service (DDoS) attack may have

been caused by a "zombie army botnet unwittingly installed on hundreds of millions of smartphones through an as yet unidentified app" [21].

- In June 2016, Softpedia reported that a botnet hiding in CCTV cameras was attacking web servers [17].
- September 2016 brought news of an army of infected IoT devices launching DDoS attacks on Brian Krebs's security blog [67], and an even larger army launching a DDoS attack on French servers [53].

In some instances, the physical distribution of IT enables new kinds of functionality. Google's Street View already requires sending out cars instrumented for video; ongoing work seeks to extend that instrumentation to things such as air quality [115]. On a larger scale, Rhett Butler at the Hawaii Institute of Geophysics and Planetology suggests using the backbone of the internet itself—the "nearly 1 million kilometers of submarine fiber optic networks...responsible for over 97 percent of international data transfer"—to sense planetary events such as earthquakes [79].

Worst-Case Scenarios: Cyber Pearl Harbor

Pearl Harbor, as most US citizens were taught in school, was the site of a surprise Japanese attack on the US Navy, which catapulted the US into World War II. In the parlance of contemporary media, the term "Pearl Harbor" has come to denote the concept of an infrastructure left completely undefended and how only a massive attack makes society take that exposure seriously.

Our society's current information infrastructure is likely full of interfaces with exploitable holes. Pundits often discuss the potential of a "cyber Pearl Harbor"—sometimes in caution, referring to the devastation that could happen if an adversary systematically exploited the holes in exactly the wrong way, but sometimes in frustration that only such a large-scale disaster would create the social will to solve these security problems.

TARGETED MALICIOUS ATTACKS IN THE IOT

As Figure 1-3 shows, researchers at SRI are already worrying about what might transpire if terrorists combined traditional kinetic methods with IoT attacks. Speculation about what might happen has even more credibility when one considers what has already happened.

THE DAILY NEWS

Thursday, April 16, 2018 THE WORLD'S FAVORITE NEWSPAPER $1.25

CHAOS AND TERROR

Cyber-Sabotaged Fire Trucks Crash Into Bombing Scene

Fire trucks responding to the bombing scene careened out of control after being sabotaged in apparent cyber attacks.

At least 20 people are dead and hundreds are injured in what appears to be a coordinated terrorist attack. Fire trucks and police units rushing down city streets to the scene of a downtown car bombing had their brakes and steering remotely disabled by cyber attacks.

Hundreds of bomb victims lay injured in the streets waiting for hours for help and many died because they did not get to a hospital in time.

According to police sources, officials have been aware for some time that emergency vehicles could be vulnerable to remote "car hacking" attacks but they did not consider it a likely terrorist threat.

Figure 1-3. Researchers at SRI speculate on the potential of augmenting traditional kinetic terrorism with IoT attacks. (Illustration by Ulf Lindqvist of SRI International, used with permission.)

The IoT has already seen vandalism with consequences. Ransomware has already been discovered in hospital computers [35] and smart TVs [16]. Analysts predict ransomware will come to medical devices themselves [34]. In spring 2016, a penetration colleague predicted it will come to all smart home appliances; in summer 2016, hackers at DefCon demonstrated a proof-of-concept for smart thermostats. On a larger scale, in early 2008, the *Register* reported [72]:

> *A Polish teenager allegedly turned the tram system in the city of Lodz into his own personal train set, triggering chaos and derailing four vehicles in the process. Twelve people were injured in one of the incidents. The 14-year-old modified a TV remote control so that it could be used to change track points.*

In 2014, the German government reported [5]:

> *A blast furnace at a German steel mill suffered "massive damage" following a cyber attack on the plant's network.... [A]ttackers used booby-trapped emails to steal logins that gave them access to the mill's control systems.*

On a related note, a penetration colleague of mine has shown me screenshots of steel mill control GUIs exposed on the open internet via VNC, noting that the GUI indicated that something was at "1,215 degrees C" and speculating on what might happen in the facility if one randomly played with the control buttons.

The IoT has also seen apparent nation-state level attacks. Perhaps the best-known of these was 2010's *Stuxnet*, malware that used a number of zero-day techniques (and even jumped across airgaps via USB thumb drives) to seek out systems connected to particular Siemens programmable logic controllers (PLCs) wired to centrifuges at Iran's uranium enrichment facilities (e.g., [69, 70]). Widely believed to have been developed and deployed by the US National Security Agency (NSA), perhaps in collaboration with Israel, Stuxnet was designed to cause enough centrifuge failures to slow down Iran's alleged weapons program —but not enough to alert Iran that sabotage might be occurring. In June 2016 a "Stuxnet copycat," also appearing to target specific installations of particular Siemens industrial control systems, was discovered, although its full picture remains unknown [23].

December 2015 brought another high-profile incident: a nation-state level actor, widely believed to be Russia, used a variety of cyber attacks to bring down the power grid in the Ukraine (e.g., [27]). Although no one particular technique here was groundbreaking, the overall coordination and scope was impressive—as was the fact that initial steps of the attack had occurred at least nine months earlier.

Where to Go Next

The IoT will continue to grow exponentially and evolve in remarkable ways. As it does, we must acknowledge that we need to prepare for and, where possible, take actions to avoid certain "fundamental truths."

First, although some vendors will try to push top-down ecosystems, the IoT will probably grow organically, a global mashup of heterogeneous components with no top-down set of principles determining its emergent behavior. This lack of control might help segment security problems at a macro scale but will be a disturbing reality for any entity that would prefer to centrally control the IoT as a large, intelligent, interconnected network. In particular, we should expect abandoned or otherwise legacy segments of today's IoT to have unanticipated interactions with and impact on the internet of tomorrow, like buried drums of highly toxic cyberwaste.

The creators of the IoT are only human and tend to replicate components at every opportunity. Industry segments are rooted in system designers' tendency to apply their favorite tools across the problem space. Common hardware and firmware libraries will show up on the IoT in surprising places; we can expect to see smart snowboards, thermostats, lightbulbs, and scientific instruments using the same connected microcontrollers and firmware. This hidden homogeneity can be bad, because just about anything having a particular "genetic" vulnerability might be compromised. However, a systematic homogeneity might also be beneficial if well-designed and inherently safe subsystems—those having the right "IoT DNA"—are widely adopted.

Consumers care little about testing regimes, product recalls, and other measures enacted in their best interest. IoT watchdog groups might start testing for compliance against a set of IoT safety standards, and governments might impose IoT safety regulations and dictate recalls, but we can expect consumers to purchase and deploy substandard devices. The IoT industry could introduce safety-assuring protocols—for instance, applying blockchains for IoT messaging. However, providers and customers will surely look elsewhere if such measures raise costs without adding significant and obvious value.

Finally, consumers' hunger for the latest and greatest might save their IoT, but enterprises' conservatism could break theirs. We might be able to exploit consumers' inherent desire for newer, better, faster to promote the ecosystem's health, at least among the consumer-facing IoT segments. We can expect today's IoT devices to get old fast, even without producers intentionally designing them to go obsolete quickly. Bad actors might get adopted quickly, but they might also fade away quickly.

WHAT DO WE DO?

It's tempting to repeat the old joke about the patient telling the doctor, "It hurts when I do this," to which the doctor replies, "Then don't do that." History shows that we keep building and deploying IT systems that contain serious vulnerabilities, which we later try to patch before too much damage is done. If the IoT's scale and distribution make this "solution" impractical, then maybe we can just start building systems without the vulnerabilities!

Unfortunately, it's not at all realistic to assume that, starting today, we'll suddenly start doing things much better. We need some game-changers: maybe a new programming language, a new approach to highly reliable input validation (e.g., [93]), or a way to use massively parallel multicore cloud computing to thoroughly fuzz-test and formally verify.

Another approach might be a new way to structure systems to mitigate damage when they're compromised. Instead of a smart grid, perhaps we need a "dumb grid": well-tested commodity operating systems, compilers, languages, and such that are modular so that developers can break off unneeded pieces. Or maybe we can use the extra cores from Moore's Law (discussed in Chapter 2) to make each IoT system multicultural: *N* distinct OSs and implementations, which aren't likely to be vulnerable in the same way at the same time.

Biology tells us that one of the problems with cancer is when the telomeres mechanism, which limits the number of times a cell can divide, stops working, allowing cell growth to run rampant. Maybe we can mitigate the problem of unlatched and forgotten IoT systems by building in a similar aging mechanism: after enough time (or enough time without patching), they automatically stop working. Of course, this could be dangerous as well. Perhaps instead, for each kind of IoT node, we can define a safe, inert "dumb" state to which it reverts after a time. (One wonders what consumers would think of this feature.)

These are just a few ideas. If we want the IoT to give us a safe and healthy future, we have our work cut out for us.

WHAT COMES NEXT

The rest of this book focuses on how the emergence of the IoT will bring many changes to how we, as a society, think about computing and manage its risks. Chapter 2 will survey some example IoT architectures. Chapter 3 will consider several IoT application areas where earlier-generation computing added smartness already, with mixed results. Chapter 4 will consider how the standard "design patterns for insecurity" that plague the IoC may surface—and be mitigated—in the IoT. Chapter 5 will examine the challenges of identity and authentication when scaled up to the IoT. Chapter 6, Chapter 7, and Chapter 8 consider privacy, economic, and legal issues, respectively. Chapter 9 considers the impact of the IoT on the "digital divide." Chapter 10 then concludes with some thoughts on the future of humans and machines, in this brave new internet.[4]

Works Cited

1. AFP, "Piege dans sa voiture, il meurt deshydrate," *Libération*, August 25, 2011.
2. AFP, "Hackers use 'smart' refrigerator to send 750,000 virus-laced emails," *Raw Story*, January 17, 2014.
3. M. Anderson, "US gas pump hacked with 'Anonymous' tagline," *The Stack*, February 11, 2015.
4. AP, "Crashed computer traps Thai politician," *Daily Aardvark*, May 14, 2003.
5. K. Ashton, "That 'Internet of Things' thing," *RFID Journal*, June 22, 2009.
6. A. Banafa, "Fog computing is vital for a successful Internet of Things (IoT)," *LinkedIn Pulse*, June 15, 2015.
7. BBC, "Hack attack causes 'massive damage' at steel works," *BBC News*, December 22, 2014.
8. K. Bora, "Volkswagen German plant accident: Robot grabs, crushes man to death," *International Business Times*, July 2, 2015.
9. C. Boutin, "Lack of effective timing signals could hamper 'Internet of Things' development," *NIST Tech Beat*, March 19, 2015.
10. W. Burleson and others, "Design Challenges for Secure Implantable Medical Device," in *Proceedings of the 49th ACM/EDAC/IEEE Design Automation Conference*, 2012.
11. R. E. Calem, "Connected car security," *CTA News*, May 17, 2016.
12. D. Chechik, S. Kenin, and R. Kogan, "Angler takes malvertising to new heights," *SpiderLabs Blog*, March 14, 2016.

4 Parts of "Worst-Case Scenarios: Cyber Love Canal" on page 1, "Worst-Case Scenarios: Cyber Pearl Harbor" on page 23, and "Where to Go Next" on page 25 are adapted from portions of my paper [96] and used with permission.

13. Check Point, "Facebook MaliciousChat," *Check Point Blog*, June 7, 2016.

14. S. Checkoway and others, "Comprehensive experimental analyses of automotive attack surfaces," in *Proceedings of the 20th USENIX Security Symposium*, 2011.

15. C. Cimpanu, "Oracle settles charges regarding fake Java security update," *Softpedia*, December 22, 2015.

16. C. Cimpanu, "Ransomware on your TV, get ready, it's coming," *Softpedia*, November 25, 2015.

17. C. Cimpanu, "A massive botnet of CCTV cameras involved in ferocious DDoS attacks," *Softpedia*, June 27, 2016.

18. C. Cimpanu, "ASUS delivers BIOS and UEFI updates over HTTP with no verification," *Softpedia*, June 5, 2016.

19. C. Cimpanu, "Medical equipment crashes during heart procedure because of antivirus scan," *Softpedia*, May 3, 2016.

20. C. Cimpanu, "Windows zero-day affecting all OS versions on sale for $90,000," *Softpedia*, May 31, 2016.

21. L. Constantin, "Attackers use email spam to infect point-of-sale terminals with new malware," *IT world*, May 25, 2015.

22. L. Constantin, "Cyberattack grounds planes in Poland," *IT world*, June 22, 2015.

23. J. Cox, "There's a Stuxnet copycat, and we have no idea where it came from," *Motherboard*, June 2, 2016.

24. J. Crook, "SmartThings and Samsung team up to make your TV a smart home hub," *TechCrunch*, December 29, 2015.

25. A. Cuthbertson, "Massive DDoS attack on core internet servers was 'zombie army' botnet from popular smartphone app," December 11, 2015.

26. Difference Engine, "Deus ex vehiculum," *The Economist*, June 23, 2015.

27. E-ISAC, *Analysis of the Cyber Attack on the Ukrainian Power Grid*. SANS Industrial Control Systems, March 18, 2016.

28. T. Eden, "The absolute horror of WiFi light switches," *Terence Eden's Blog: Mobiles, Shakespeare, Politics, Usability, Security*, March 2, 2016.

29. J. Epstein, "Risks of online test taking," *The Risks Digest*, May 21, 2015.

30. A. C. Estes, "Why is my smart home so fucking dumb?," *Gizmodo*, February 12, 2015.

31. S. Fink, *Five Days at Memorial*. Deckle Edge, 2013.

32. M. Finnegan, "Toyota recalls 1.9m Prius cars due to software fault," *Computerworld UK*, February 2014.

33. D. Fisher, "Gone in less than a second," *ThreatPost*, August 6, 2015.

34. D. Fisher, "What's on TV tonight? Ransomware," *On the Wire*, June 13, 2016.

35. T. Fox-Brewster, "As ransomware crisis explodes, Hollywood hospital coughs up $17,000 in Bitcoin," *Forbes*, February 18, 2016.

36. T. Fox-Brewster, "White hat hackers hit 12 American hospitals to prove patient life 'Extremely Vulnerable'," *Forbes*, February 23, 2016.

37. R. Franck, "Beveiliging sneltram valt niet te vertrouwen," *Algemeen Dagblad*, April 3, 2016.

38. S. Gallagher, "Report: Airbus transport crash caused by 'wipe' of critical engine control data," *Ars Technica*, June 10, 2015.

39. N. Gamer, "Vulnerabilities on SoC-powered Android devices have implications for the IoT," *Trend Micro Industry News*, March 14, 2016.

40. M. Garrett, "I stayed in a hotel with Android lightswitches and it was just as bad as you'd imagine," *mjg59's journal*, March 11, 2016.

41. S. Gibbs, "Security researchers hack a car and apply the brakes via text," *The Guardian*, August 12, 2015.

42. S. Gibbs, "Windows 10 automatically installs without permission, complain users," *The Guardian*, March 15, 2016.

43. A. Goard, "Thief steals $15,000 bike in Sausalito with tap of hand: Police," *NBC Bay Area*, February 27, 2015.

44. J. Golson, "Many Lexus navigation systems bricked by over-the-air software update," *The Verge*, June 7, 2016.

45. D. Goodin, "Rise of 'forever day' bugs in industrial systems threatens critical infrastructure," *Ars Technica*, April 9, 2012.

46. D. Goodin, "Foul-mouthed worm takes control of wireless ISPs around the globe," *Ars Technica*, May 19, 2016.
47. C. Graeber, "How a serial-killing night nurse hacked hospital drug protocol," *Wired*, April 29, 2013.
48. A. Greenberg, "Chrysler catches flak for patching hack via mailed USB," *Wired*, September 3, 2015.
49. A. Greenberg, "Hackers remotely kill a Jeep on the highway—with me in it," *Wired*, July 21, 2015.
50. S. Greengard, *The Internet of Things*. MIT Press, 2015.
51. S. Harris, "Engineers work to prevent electric vehicle charging from overloading grids," *The Engineer*, April 13, 2015.
52. hdmoore, "The Internet of gas station tank gauges," *Rapid7 Community*, January 22, 2015.
53. B. Hill, "Latest IoT DDoS attack dwarfs Krebs takedown at nearly 1Tbps driven by 150K devices," *Hot Hardware*, September 27, 2016.
54. S. P. Holland and others, "Are there environmental benefits from driving electric vehicles?," *American Economic Review*, December 2016.
55. I. Johnston, "Smartwatches that allow pupils to *cheat* in exams for sale on Amazon," *The Independent*, March 3, 2016.
56. R. V. Jones, *Most Secret War*. Penguin, 2009.
57. D. Kemp, C. Czub, and M. Davidov, *Out-of-Box Exploitation: A Security Analysis of OEM Updaters*. Duo Security, May 31, 2016.
58. A. Kessler, "Fiat Chrysler issues recall over hacking," *The New York Times*, July 24, 2015.
59. O. Khazan, "How to fake your workout," *The Atlantic*, September 28, 2015.
60. J. Kirk, "How encryption keys could be stolen by your lunch," *Computerworld*, June 22, 2015.
61. Z. Kleinman, "Microsoft accused of Windows 10 upgrade 'nasty trick'," *BBC News*, May 24, 2016.
62. W. Knight, "Baidu's self-driving car takes on Beijing traffic," *MIT Technology Review*, December 10, 2015.

63. G. Kolata, "Of fact, fiction and Cheney's defibrillator," *The New York Times*, October 27, 2013.

64. P. Koopman, J. Black, and T. Maxino, "Position paper: Deeply embedded survivability," in *Proceedings of the ARO Planning Workshop on Embedded Systems and Network Security*, 2007.

65. K. Koscher and others, "Experimental security analysis of a modern automobile," in *Proceedings of the IEEE Symposium on Security and Privacy*, 2010.

66. B. Krebs, "Lizard stresser runs on hacked home routers," *Krebs on Security*, January 9, 2015.

67. B. Krebs, "KrebsOnSecurity hit with record DDoS," *Krebs on Security*, September 21, 2016.

68. B. Krebs, "Target hackers broke in via HVAC company," *Krebs on Security*, February 5, 2014.

69. D. Kushner, "The real story of Stuxnet," *IEEE Spectrum*, February 26, 2013.

70. R. Langner, "Stuxnet's secret twin," *Foriegn Policy*, November 19, 2013.

71. S. Levin and N. Woolf, "Tesla driver killed while using Autopilot was watching Harry Potter, witness says," *The Guardian*, July 1, 2016.

72. J. Leyden, "Polish teen derails tram after hacking train network," *The Register*, January 11, 2008.

73. N. Lomas, "The FTC warns Internet of Things businesses to bake in privacy and security," *TechCrunch*, January 8, 2015.

74. P. Longeray, "Windows 3.1 is still alive, and it just killed a French airport," *Vice News*, November 13, 2015.

75. J. Lowy, "Will robot cars drive traffic congestion off a cliff?," *AP: The Big Story*, May 16, 2016.

76. A. MacGregor, "BMW patches security flaw affecting over 2 million vehicles," *The Stack*, February 2, 2015.

77. J. Markoff and C. C. Miller, "As robotics advances, worries of killer robots rise," *The New York Times*, June 16, 2014.

78. L. Mearian, "With $15 in Radio Shack parts, 14-year-old hacks a car," *Computerworld*, February 20, 2015.

79. D. Oberhaus, "The backbone of the internet could detect earthquakes, but no one's using it," *Motherboard*, August 11, 2015.

80. K. Palani, E. Holt, and S. W. Smith, "Invisible and forgotten: Zero-day blooms in the IoT," in *Proceedings of the 1st IEEE PerCom Workshop on Security, Privacy, and Trust in the IoT*, 2016.

81. A. Pasick, "An iPad app glitch grounded several dozen American Airlines planes," *Quartz*, April 28, 2015.

82. Paul, "Research: IoT hubs expose connected homes to hackers," *The Security Ledger*, April 7, 2015.

83. Paul, "X-rays behaving badly: Devices give malware foothold on hospital networks," *The Security Ledger*, June 8, 2015.

84. N. Perlroth, "Traffic hacking: Caution light is on," *The New York Times*, June 10, 2015.

85. S. Peterson and P. Faramarzi, "Exclusive: Iran hijacked US drone, says Iranian engineer," *The Christian Science Monitor*, December 15, 2011.

86. R. Price, "Google's parent company is deliberately disabling some of its customers' old smart-home devices," *Business Insider*, April 4, 2016.

87. M. Reel and J. Robertson, "It's way too easy to hack the hospital," *Bloomberg Businessweek*, November 2015.

88. D. Roberts, "New study: Fully automating self-driving cars could actually be worse for carbon emissions," *Vox*, February 27, 2016.

89. J. Rogers, "LA traffic is getting worse and people are blaming the shortcut app Waze," *The Associated Press*, December 14, 2014.

90. I. Rouf and others, "Security and privacy vulnerabilities of in-car wireless networks: A tire pressure monitoring system case study," in *Proceedings of the 20th USENIX Security Symposium*, 2010.

91. B. Salsberg, "Mass. train leaves station without driver," *Burlington Free Press*, December 10, 2015.

92. A. San Juan, "Texas man, dog die after being trapped in Corvette," *KHOU*, June 10, 2015.

93. L. Sassaman and others, "Security applications of formal language theory," *IEEE Systems Journal*, September 2013.

94. R. Singel, "Industrial control systems killed once and will kill again, experts warn," *Wired*, April 9, 2008.

95. C. Smith, *The Car Hacker's Handbook*. No Starch Press, 2016.

96. S. W. Smith and J. S. Erickson, "Never mind Pearl Harbor—What about a cyber Love Canal?," *IEEE Security and Privacy*, March/April 2015.

97. T. Smith, "Hacker jailed for revenge sewage attacks," *The Register*, October 31, 2001.

98. A. Southall, "Technical problem suspends flights along East Coast," *The New York Times*, August 15, 2015.

99. D. Spaar, "Beemer, open thyself!—Security vulnerabilities in BMW's ConnectedDrive," *c't*, May 2, 2015.

100. L. Stack, "J.F.K. Computer glitch wreaks havoc on air passengers," *The New York Times*, May 30, 2016.

101. B. Steele, "Windows 10 update message interrupts live weather report," *Engadget*, April 28, 2016.

102. I. Thomson, "Even in remotest Africa, Windows 10 nagware ruins your day: Update burns satellite link cash," *The Register*, June 3, 2016.

103. A. Toor, "Heinz ketchup bottle QR code leads to hardcore porn site," *The Verge*, June 19, 2015.

104. Tribune wire reports, "Drones banned from World Cup skiing after one nearly falls on race," *Chicago Tribune*, December 23, 2015.

105. Z. Tufekci, "Why 'smart' objects may be a dumb idea," *The New York Times*, August 10, 2015.

106. J. Valentino-Devries, "Rarely patched software bugs in home routers cripple security," *The Wall Street Journal*, January 6, 2016.

107. C. Vallance, "Car hack uses digital-radio broadcasts to seize control," *BBC News*, July 22, 2015.

108. J. Voelcker, "Tesla Model S hacked in low-speed driving; Patch issued, details tomorrow," *Green Car Reports*, August 6, 2015.

109. G. Wang and others, "Defending against Sybil devices in crowdsourced mapping services," in *MobiSys '16, Proceedings of the 14th Annual International Conference on Mobile Systems, Applications, and Services*, 2016.

110. P. Wayner, *Future Ride v2*. CreateSpace, 2015.

111. R. Wiedeman, "The Big Hack: The day cars droves themselves into walls and the hospitals froze," *New York*, June 19, 2016.

112. K. Wilhoit, *The SCADA That Didn't Cry Wolf: Who's Really Attacking Your ICS Equipment? (Part 2)*. Trend Micro Research Paper, 2013.

113. K. Wilhoit, *Who's Really Attacking Your ICS Equipment?* Trend Micro Research Paper, 2013.

114. K. Wilhoit and S. Hilt, *The GasPot Experiment: Unexamined Perils in Using Gas-Tank-Monitoring Systems*. TrendLabs, 2015.

115. C. Wood, "Google to measure air quality through Street View," *Gizmag*, July 30, 2015.

116. K. Zetter, "It's insanely easy to hack hospital equipment," *Wired*, April 25, 2014.

117. K. Zetter, "All U.S. United flights grounded over mysterious problem," *Wired*, June 2, 2015.

118. K. Zetter, "Feds say that banned researcher commandeered plane," *Wired*, May 15, 2015.

119. V. Zhang, "High-profile mobile apps at risk due to three-year-old vulnerability," *TrendLabs Security Intelligence Blog*, December 8, 2015.

120. Y. Zitun and E. B. Kimon, "IDF investigation finds soldiers in Qalandiya acted appropriately," *YNet News*, March 1, 2016.

121. Z. Zorz, "65,000+ Land Rovers recalled due to software bug," *Help Net Security*, July 14, 2015.

122. Z. Zorz, "Cheap web cams can open permanent, difficult-to-spot backdoors into networks," *Help Net Security*, January 14, 2016.

123. Z. Zorz, "Researchers hack the Mitsubishi Outlander SUV, shut off alarm remotely," *Help Net Security*, June 6, 2016.

124. Z. Zorz, "1,400+ vulnerabilities found in automated medical supply system," *Help Net Security*, March 30, 2016.

Examples and Building Blocks

To guide the rest of the discussion, I'll start by presenting a basic framework for the IoT: a population of (physically) small computing devices, tied to the physical world with sensors and actuators, possibly communicating with one another and with backend systems.

Computing Devices

To make the book self-contained for the reader without a detailed computer science background, this section discusses the basics of computer systems (central processing unit, memory, software structure) and how "embedded systems" differ from regular ones.

BASIC ELEMENTS

Generally speaking, computing devices consist of a central processing unit (CPU), memory, and some kind of input/output (I/O). The memory typically includes some kind of nonvolatile storage (so there's a program to run when the machine first powers up) and some kind of slower but more efficient backing store for longer-term storage. In a simpler universe, the CPU would fetch an instruction from memory, figure out what the instruction means, carry it out, and then go on to the next one. "Carrying out" the instruction would involve reading or writing to memory or an I/O device or an internal CPU register, and/or doing some mathematical operation. However, the universe is not that simple anymore. The field traded simplicity for speed (as I have to cover in the "dirty tricks" section of my architecture course): pieces of many instructions may be carried out at the same time and may be executed out of order or speculatively; parts of memory may be tucked away in smaller but faster caches.

I/O typically includes some way of interacting with a human (e.g., keyboard and display) and some kind of networking to enable interaction with other computing devices.

The hardware also requires software: the BIOS to run at boot time, application programs, and an operating system to provide a clean and hopefully protected environment for the applications.

MOORE'S LAW

An important part of understanding the long-term progression of computing is *Moore's Law*. Despite its official-sounding name, it's not actually a "law" (like those of Newton), and it's not as well defined as one might like: ask two computing professionals, and you will likely hear two different versions.

Moore's Law refers to a pattern observed by Intel cofounder Gordon Moore decades ago, at the dawn of the modern computing era. Every *N* years, the complexity of integrated circuits—measured in various ways, such as transistor density—doubles (see Figure 2-1). The pattern of growth has continued to the present day, despite regularly occurring reports that some fundamental physical or engineering limit has been reached and it cannot possibly be sustained. It's also generally acknowledged that the continued existence of this pattern is in some sense a self-fulfilling prophesy: it can take several years to design the next-generation microprocessor, and the industry may start the design assuming a fabrication ability that that does not yet exist but is forecast to exist by the time the design is done. (Using a simile from American football, some[1] have quipped it's "like throwing a Hail Mary pass, then running down the field and catching it.")

1 Sometimes, this quote is attributed to Dave Patterson, professor in Computer Science at UC Berkeley.

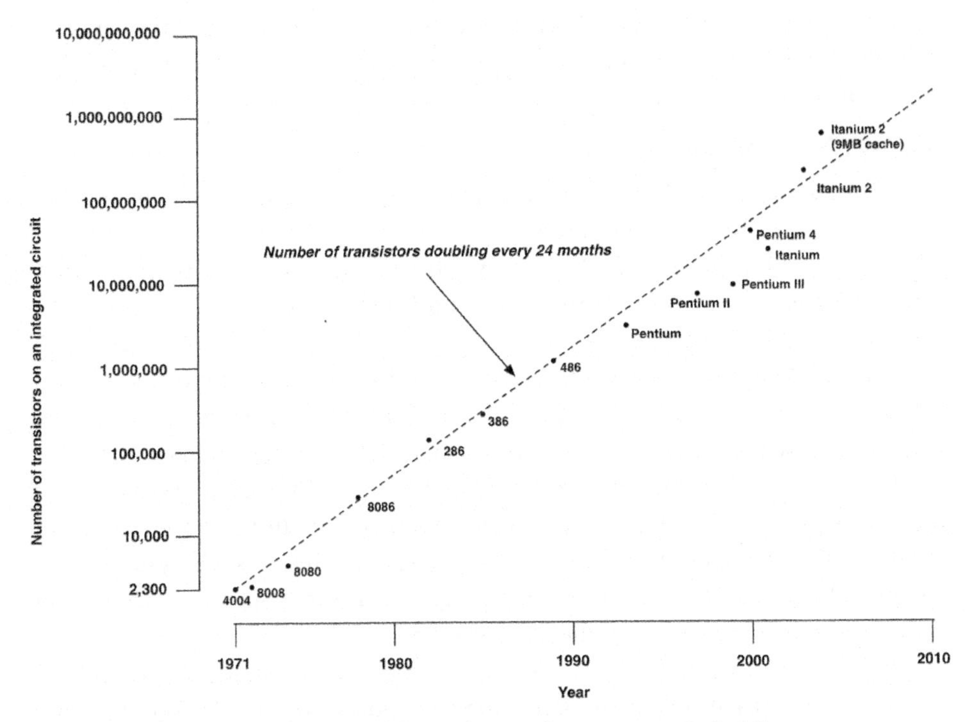

Figure 2-1. Computation has followed the prediction of Moore's "Law": doubling every two years, power increases exponentially. (Adapted from WikiMedia Commons, public domain.)

When we step back and look at the progress of computing, the exponential growth characterized by Moore's Law has several implications. The "current"-generation CPU keeps getting exponentially more complex. As one consequence, computing applications that might previously have seemed silly and impractical (because they would take far too much computing time and memory) become practical and commonplace. (The things done today by ubiquitous smartphones would have been dismissed as too crazy to consider had they been proposed as research projects in 1987—I was there, and I was one of the ones doing the dismissing.) We also get the flip side of increased complexity: large systems (especially software systems) are harder to get right and can have surprising behaviors and error cases.

Another consequence is that things get cheaper—particularly as *previous*-generation processors become repurposed in embedded devices. As an *XKCD* cartoon observed, even if you stay two steps behind the technology curve, you still see the same exponential growth—only it's much cheaper! In the classroom, I've taken to borrowing a colleague's trick: comparing the surprising prevalence of previous-generation processors to the surprising prevalence of "lesser" species, such as insects.

HOW IOT SYSTEMS DIFFER

By definition, most IoT devices will be *embedded systems*: computers embedded in things that don't look like computers.

If one were to ask how embedded systems differ from standard computers, the initial answer would be, "they're much smaller and don't do as much." To save money and because the systems might not have to do very much, the physical size and computational power might be limited. Systems that must run on batteries will also be designed to reduce power consumption. Systems that do not work directly with humans may skip the human I/O devices; systems without the need to interact with other systems may skip the networking. Systems that do not do very much may also skip the nonvolatile backing store (if there's not much to store) or use semiconductor flash memory instead of disks; such systems may additionally have an OS that does not support much multitasking (if there's not much to be done), or even skip the OS altogether.

However, thanks to Moore's Law, this initial answer is starting to be less true, because "small, cheap" computational engines are also becoming more powerful. When it comes to prognosticating the future IoT, Moore's Law has two somewhat different implications:

1. Thanks to the exponential curve, many IoT devices deployed in the future will be powerful, strong computational engines, by today's standards. Indeed, we're already seeing this creeping functionality and power in IoT devices, for instance, full Linux installations and network support and built-in web servers. (I recently discovered my Kindle reader has a web browser!)

2. IoT devices embedded in long-lived "things" will then fall behind the curve; by the standards current *N* years from now, the computing devices embedded in older refrigerators, cars, and bridges will likely seem quaint and archaic.

However, note that I opened this section with the qualifier "most." A computer is a digital circuit that does things (certain inputs trigger certain outputs and cause the internal state to change), but structured in a particular way: one piece executing instructions coded as binary numbers stored in another piece. One can build digital circuits to provide such behaviors without the full Turing generality of a computing engine: indeed, this why Dartmouth's Engineering School offers both a "Digital Design" course and a "Microprocessors" course. My soapbox conjecture is that, due to the incredible flexibility and reusability of standard components—as well as the computational nature of many of the tasks we want them to do—the IoT will mainly consist of computers, rather than noncomputational engines.

Another way embedded systems can differ from traditional computers is in their lifetime: a computer embedded in some other kind of thing needs to live as long as that thing. Greeting cards that play "Happy Birthday" don't live very long, but automobiles and appliances may last decades, and bridges and electrical grid equipment even longer. Chapter 4 will discuss some of the relevant security issues arising from computational power, exposed interfaces, software complexity, and long lifetimes.

Architectures for an IoT

The preceding discussion frames how we might think of future IoT systems. To start with, they will be embedded computing systems that are likely to be fairly rich, by standards current not too long before the systems get created.

Just to make things tangible, Figure 2-2 shows an already-old IoT module with a 500 MHz processor, half a gigabyte of RAM, flash and an SD card for nonvolatile storage, and Ethernet. Not that many years ago, these would have been the specs of a nice desktop device—and when they were, did anyone imagine that engineering would so quickly advance that it would soon be economically feasible to put physically tiny versions in household appliances?

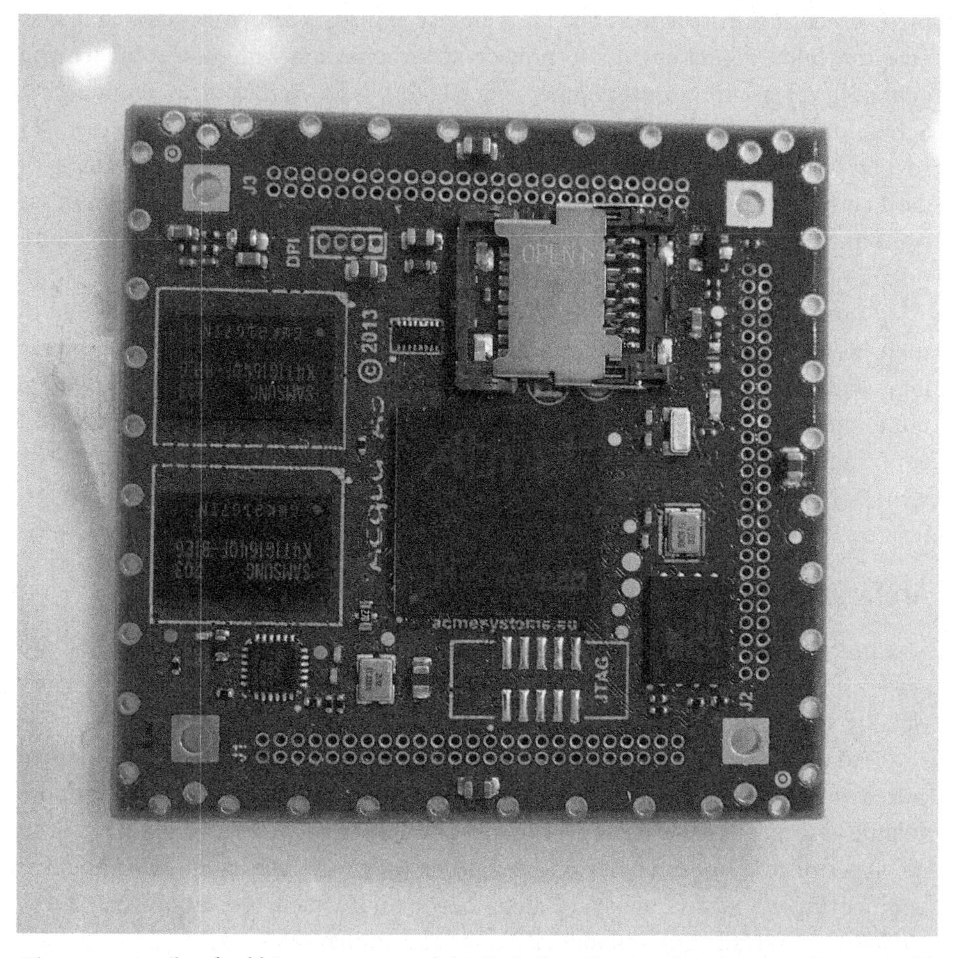

Figure 2-2. An already-old "system on a module" for IoT applications has the specs of what would have been a nice desktop platform not too long ago. (From Wikimedia Commons, public domain.)

Other parameters in an informal taxonomy of IoT architecture include connection to other computers, connection to the physical world, and connection to big backend servers. Let's consider each issue in turn.

CONNECTION TO OTHER COMPUTERS

Will IoT devices be networked? Yes, it seems a bit ironic that something we're calling the *Internet* of Things may in fact not involve networking, but the working definition we've been using—adding computational devices to everything around us—allows for that possibility. (Indeed, I recall a very elegant and effective "smart home appliance" pilot where computation added to the appliance could sense when the electric grid was stressed and adjust its internal operation appropriately without ever needing to talk to anything else.)

For devices that are networked, what is the physical medium of communication? The natural choices here are wire (e.g., the Ethernet connection in Figure 2-2) or radio. (To make things tangible, Figure 2-3 shows some inch-sized radio modules for IoT applications.) In the research world, a colleague of mine is even exploring the potential of using light as a medium, even in a dark room.

Figure 2-3. Example of small, inexpensive radio modules for IoT applications. (From Wikimedia Commons, Mark Fickett.)

When it comes to radio, it's important to remember that electromagnetic radiation is not magic. Depending on the choice of power and wavelength (proportional to frequency), textbook-level analysis predicts signals from wireless networking will travel anywhere from a few centimeters to many kilometers. However, textbooks are not the same as the real world; an old-time engineer I worked with often ranted that young computer scientists would believe that if the spec said signals would travel 10 meters, then any device within 10 meters would receive them perfectly but any device beyond 10 meters would hear nothing. As anyone who has tried to make a cellphone call in Vermont or northern New Hampshire knows, it's not that simple.

When discussing radio-based networking, it's tempting to use the term "WiFi." However, this is somewhat akin to using the term "Kleenex" for a facial tissue: the WiFi Alliance established the term for wireless networking conforming to certain (evolving) IEEE standards, but there are other approaches as well, such as Bluetooth (an example of closer-range communication) and *ZigBee* (which has been seeing deployment in industrial control settings). Many in the industry eagerly await the coming of 5G wireless networking, heralded to provide vastly greater speeds and increased reliability.

One interesting consequence of new wireless stacks and the shrinking packaging from Moore's Law is that, initially, it can be hard to probe the wireless interface for security holes due to unplanned input (see "Instance: Failure of Input Validation" on page 72) because the interface is buried inside a larger module; the article "Api-do: Tools for exploring the wireless attack surface in smart meters" discusses some of my lab's work in this space [5].

Beyond the physical medium itself, there's the question of the protocols used to communicate. Again, "internet" in the term "Internet of Things" suggests to some that the IoT must, of necessity, use some flavor of the standard Internet Protocol (IP) stack. As with any engineering decision, there are advantages and disadvantages to this design choice—on the one hand, it's nice to be interoperable with the known universe (i.e., the Internet of Computers). But on the other hand, incorporating the IP stack may incur performance and complexity costs that don't fit within the constraints of the real-world application. daCosta [2] makes an excellent case for this latter point. (And, as Chapter 5 will explore, many IoT application scenarios may not require the fully general communication pattern of the internet.)

In the Internet Protocol, each device receives a unique address. The emerging IPv6 (which has been in the process of succeeding the current IPv4 for a decade or so) is often touted as a special enabler of the IoT because IPv4 only allowed for 2^{32} addresses (approximately 4 billion), whereas IPv6 allows for 2^{132} —a factor of a trillion billion billion more.

CONNECTION TO THE PHYSICAL WORLD

IoT devices will interact with the physical world around them.

Because most humans have experience with the IoC, we are used to the idea of keys and buttons being input channels to computing devices, and screens and displays and blinking lights being output channels. Because computers are electronic devices, it's also not a stretch to think of electrical actions as I/O, for example, a computer embedded in an air conditioner turning on a fan. However, IoT devices can also employ components—sensors and actuators—enabling more direct kinetic input and output.

Fu and Jaradat's (free) report [4], prepared for the Maryland State Highway Administration, shows many nice examples of such sensors—strain gauges and displacement transducers (Figure 2-4), inclinometers, GPS sensors, accelerometers—and their use in letting smart devices monitor the state of physical infrastructure such as bridges and dikes. *Stepper motors* (Figure 2-5) are just one example of an actuator enabling a computer to trigger precisely controlled physical movement.

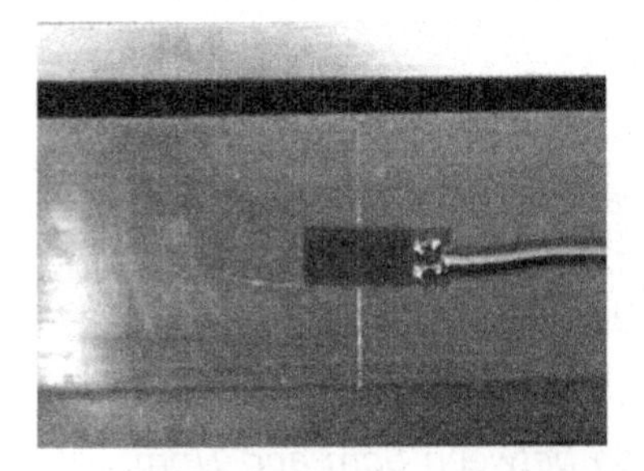

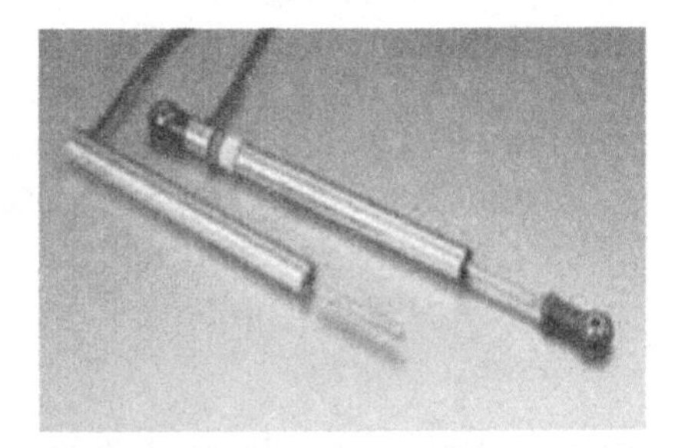

Figure 2-4. Sensors such as strain gauges (left) and displacement transducers (right) may be used for the smartening of physical infrastructure, such as bridges. (From [4], used with permission.)

Figure 2-5. An IoT device may use actuators such as this stepper motor to change the state of physical infrastructure in which it's embedded. (From Wikimedia Commons, Dolly1010 - Vlastní fotografie.)

When I taught my "Risks of the IoT to Society (RIOTS)" class at Dartmouth, I launched the students with Raspberry Pis and stacks of Sunfounder sensor/actuator modules. Figure 2-6 shows one example of what happened, described by the students as follows [3]:

> *The PiPot greatly simplifies the process of growing indoor plants by automating watering and light exposure. The Pi is attached to a moisture sensor, a plant grow lamp, and a watering system. The watering system uses a storage bag for water, connected to a tube that goes through a solenoid valve....*
>
> *The grow lamp is programmed to turn on between 5pm and 11pm. However, weather is taken into consideration—this model is for plants that are indoors, but are near windows and receive some natural sunlight. Between 7am and 5pm, the Pi will check to see if the weather outside is "fair" or "clear"....*
>
> *Every hour, the attached moisture sensor will check moisture levels in the soil. If moisture is below the desired threshold, in our case under 150, the solenoid valve will open to water the plant for three seconds. An hour later, the sensor checks again. This cycle of lighting and watering continues every day until the user ends the program.*

Figure 2-6. One student project in my "Sophomore Summer" IoT class used moisture sensors and daylight and weather information to control watering and lights for indoor plants. (Ke Deng and Dylan Scandinaro, used with permission.)

THE BACKEND

As discussed earlier, Moore's "Law" is bringing exponential increases in semiconductor density; leading to increases in the power and storage ability of computing; in the complexity of what becomes "feasible" to realize in computer software; and in affordability. However, these trends affect not just small devices deployed in the IoT but also the big devices in the IoC. The computing industry has seen consequences in many areas: Cloud computing:: Giant (but virtual and remote) servers available for inexpensive rental. Big data:: Massive amounts of available data on which to compute. Big data analytics:: Increases in the sophistication of what can be feasibily calculated from these large data sets. Deep learning:: Increases in the sophistication of one particular family of approaches.

(A deeper explication of analytics and deep learning is beyond the scope of this book, but for an introduction, see Chapter 17 in my earlier security textbook [8].)

Other industries have been quick to exploit these consequences. To quote a McKinsey report [6]:

> *The data that companies and individuals are producing and storing [in one year] is equivalent to filling more than 60,000 US Libraries of Congress....*
>
> *Big data levers offer significant potential for improving productivity at the level of individual companies.... For instance, Tesco's loyalty program generates a tremendous amount of customer data that the company mines to inform decisions from promotions to strategic segmentation of customers. Amazon uses customer data to power its recommendation engine "you may also like ..." based on a type of predictive modeling technique called collaborative filtering.... Progressive Insurance and Capital One are both known for conducting experiments to segment their customers systematically and effectively and to tailor product offers accordingly.*

These transformations of "big" computing are relevant to the tiny devices in the IoT because they can be tied together. Connecting the IoT to the IoC lets the IoT take advantage of this big data backend, both by providing new sources of "big data" and by taking actions driven by the big data analytics. Indeed, the tie is so tight that a set of guidelines to bring security to IoT product development was recently issued—by the *Cloud* Security Alliance [1]. In their *Harvard Business Review* article [7] (discussed further in Chapter 7), Porter and Heppelmann see the IoT/IoC connection as a critical part of how the IoT will transform business (and, hence, the world). On the flip side, some lament the problem of *digital exhaust*: the fact that much of the data being generated by an instrumented world is not actually used.

Of course, the tie-in also lets us use all the fashionable buzzwords—IoT, cloud, big data, analytics—in a single sentence while keeping a straight face.

The Bigger Picture

This chapter sketched some basic building blocks for the IoT as examples. However, this sketch is just an illustrative slice of a very large and active field. For example, venture capitalist Matt Turck and colleagues have been putting together an annual "Internet of Things Landscape" showing various companies making moves in this space. The 2016 landscape (*http://bit.ly/2gSDVTD*) lists almost 200 companies active in just the "Building Blocks" domain, and another 1,000 spread over "Applications" and "Platforms and Enablement." The rather complicated

Figure 2-7 from Beecham Research provides an incomplete but representative sketch of IoT applications.

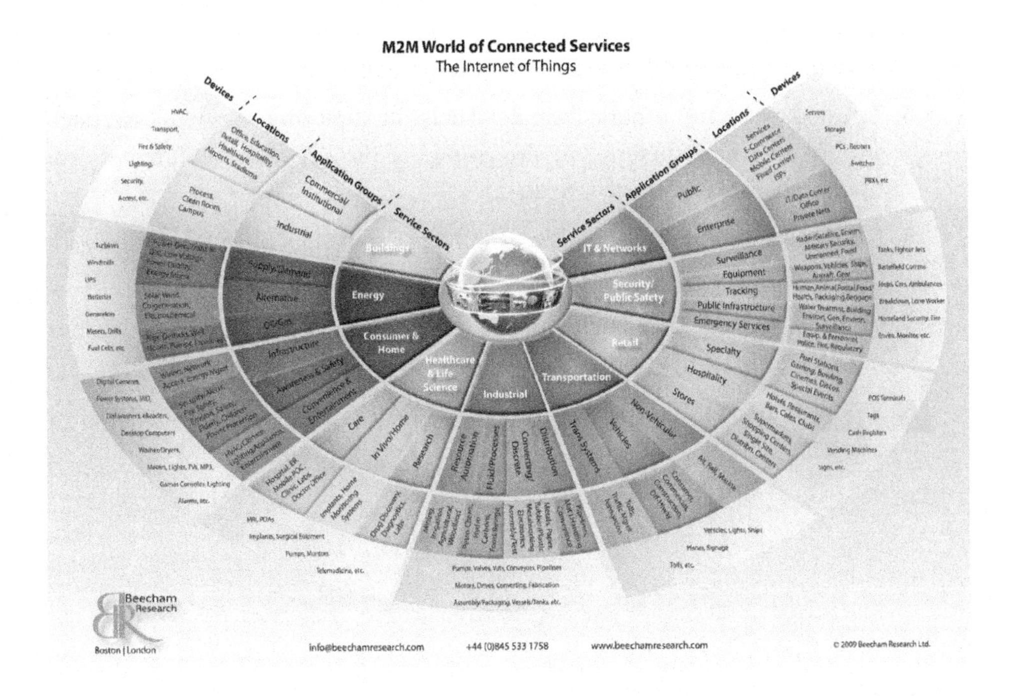

Figure 2-7. The technology market analysis firm Beecham Research draws this map of business development activity in IoT—it's from 2009 and already incomplete. (Used with permission.)

What's Next

With this basic sketch of the IoT and its IoC backend in hand, we now dive into the rest of the story.

Works Cited

1. CSA, *Future-Proofing the Connected World: 13 Steps to Developing Secure IoT Products*. Cloud Security Alliance IoT Working Group, 2016.

2. F. daCosta, *Rethinking the Internet of Things: A Scalable Approach to Connecting Everything*. Apress Media, 2013.

3. K. Deng and D. Scandinaro, *Final Project: PiPot.* Dartmouth College COSC69 Fancy Application Report, Summer 2015.

4. C. C. Fu and Y. Jaradat, *Survey and Investigation of the State-of-the-Art Remote Wireless Bridge Monitoring System.* Maryland Department of Transportation State Highway Administration, 2008.

5. T. Goodspeed and others, "Api-do: Tools for exploring the wireless attack surface in smart meters," in *Proceedings of the 45th Hawaii International Conference on System Sciences*, 2012.

6. J. Manyika and others, *Big data: The Next Frontier for Innovation, Competition, and Productivity.* McKinsey Global Institute, May 2011.

7. M. E. Porter and J. E. Heppelmann, "How smart, connected products are transforming competition," *Harvard Business Review*, November 2014.

8. S. Smith and J. Marchesini, *The Craft of System Security.* Addison-Wesley, 2008.

3

The Future Has Been Here Before

Conversations about how the IoT will "change everything" span many application domains. Three sexy ones in particular are:

Smart medicine
: Putting IT inside medical devices

Smart grid
: Putting IT everywhere in the power grid

Smart cars
: Putting IT everywhere inside cars—even replacing the driver

Discussions of the impact of the IoT—of making everything smart—usually go in two directions. In one (reminiscent of *Pollyanna* or *Candide*'s Pangloss) the vision is impossibly rosy: the IoT will make everything wonderful. Another direction, reminiscent of the Cassandra from mythology touts a dystopia (too often what listeners hear when security people speak). Nonetheless, both visions have a common thread: this future is *new*.

Although the IoT is new, interconnecting IT with physical reality (albeit on a less massive scale) is not. However, this is not the first time that information technology has taken a quantum jump in its integration with non-IT aspects of life—even for domains the IoT is penetrating. This is also not the first time that Cassandras and Pollyannas have spoken.

This chapter revisits some older incidents, not widely known outside the software engineering community, where interconnection of complex IT with physical reality had interesting results. The discussion follows the sexy application domains just listed:

- For smart medicine, what happened back in the 1980s with the simpler Therac-25, a computer-controlled radiation therapy device
- For the smart grid, what happened in the 2003 East Coast blackout with the simpler cyber infrastructure of the last decade's power grid
- For smart cars, what happened 30 years ago when the Airbus A320 brought computer-based control to passenger aircraft

In each of these cases, it's safe to say there were unexpected consequences from the IT–physical connection. Sometimes Cassandras spoke and were wrong; sometimes they didn't but should have. The past is prologue; in each case, the history sheds some light on current IoT efforts in these spaces.

Bug Background

When talking about some consequences of layering IT on top of non-IT technology, it helps to discuss a few types of bugs that can surface in large software systems (as programs glued to these internetworked things tend to be). Experienced programmers will likely recognize these patterns.

INTEGER OVERFLOW

Software design, in some sense, is a form of translation: expressing higher-level concepts of workflow and physical processes and what have you in terms of precise instructions carried out by a computing machine. As with any type of translation (or mathematical mapping from one set to another), the higher-level idea and its machine representation may not necessarily be identical.[1] As software often deals with basic manipulation of numbers, one manifestation of this translation error arises from considering questions such as "How big is an integer?" In the world of mathematics, an integer is an integer; there are an infinite number of them. However, computing machines typically represent an integer as a binary number and reserve a specific number of bits (binary digits) to represent *any* integer.

1 Chapter 10 will revisit the idea of looking at IT security and usability troubles in terms of semantic mismatches.

For example, suppose the variable `x` is intended to represent a nonnegative integer. In the programmer's mental model of what he or she is doing, `x` can have an infinite number of values, and it's always going to be true that `x + 1` will be greater than `x`. But if the computing machine represents `x` as an 8-bit value, then there are only $2^8 = 256$ possible values. An infinite number of things cannot fit into only 256 slots. In the representation machines naturally use, if `x = 255` and the program adds 1, then `x = 0`, surprisingly (unless one is familiar with modular arithmetic). The rule that adding 1 gives a larger number no longer holds.

RACE CONDITIONS

Another family of bugs arises when software may consist of two or more threads of execution happening at the same time ("concurrently" or "in parallel," depending on the mechanism). Actions by thread *A* may have dependencies on actions by thread *B* in ways obvious or subtle, and if a particular run of the system fails to respect these dependencies, things can go dramatically wrong.

As an extreme noncomputing example, suppose the job of thread *Alice* is to drive a commuter train and the job of thread *Bob* is to take the train to work. Although Alice and Bob may be two separate actors, coordination is critical: if Bob steps off the station platform a few seconds too early—just as the train is arriving—the outcome will not be good. (The correctness of the execution thus depends on a "race" between the two threads.)

In computing, the deadly interaction typically involves inadvertent concurrent use of the same data structure, and is often much more subtle. What's even more vexing is that if the programmer does not see and account for the dependency at design time, it can be very hard for testing to reveal it. Imagine that Alice's thread of execution is a large red deck of numbered playing cards, and Bob's is a large green deck. What actually happens in any particular execution of the system is a shuffling together of these two decks: the red and green cards each stay in order, but their interleaving is arbitrary. If the bad action happens only if Bob's card #475 gets placed between Alice's #242 and #243, then one might run the system many times and never see that happen—and when it does happen, one might rerun the system and see the buggy behavior disappear.

MEMORY CORRUPTION

Computers store data and code in numbered memory slots and manipulate an item by specifying the index number (*address*) of its slot. *Memory corruption* bugs occur when the program somehow gets the correspondence of address to slot wrong, and writes the wrong thing in the wrong place. For a noncomputing metaphor, think of the letter in *Hamlet* instructing that the bearer of the letter should be killed; this plan has unexpected consequences if the wrong person carries the letter.

IMPOSSIBLE SCENARIOS

Again, designing and reasoning about computer system behavior involves a semiotic mapping between the reality of what a system actually does and the mental model the programmer or analyst has of what it does. In many situations (including the bug scenarios just described), things can get lost in translation; behaviors can happen in the actual system that are impossible in the mental model. This phenomenon can make it very hard to find and fix the problems.

The future has been here before—and all of these bugs showed up.

Smart Health IT

If the IoT is defined as "layering IT on top of previously noncomputerized things," one area receiving much attention is *smart health*. Healthcare requires measuring and sensing and probing and manipulating what is quite literally a physical system: the human body. Adding a cyber component to this creates some interesting potential. For a Panglossian voice, consider this quote from Miller's textbook [12]:

> *It's a given that the Internet of Things (IoT) is going to change healthcare as we know it. By connecting together all the various medical devices in use (or soon to be introduced), healthcare gets a lot smarter real fast.... It's a medical dream come true.*

Smart health may be an exciting new avenue in the 2010s, but this future has also been here before. As a cautionary tale, we will revisit the 1980s and the *Therac-25*, a computer-controlled radiation therapy machine.

THE THERAC-25

One of the treatments for cancer is radiation: bombarding the malignant cells with focused beams of high-energy particles. The Therac-25, from Atomic Energy of Canada Limited (AECL), contained a 25 million-electron volt accelerator and could generate either electron or X-ray photon beams; the latter requires first generating a much higher-energy electron beam and then converting it. These processes—particularly the conversion to X-ray—required that various objects be physically arranged with the beam; a primary component in this arrangement was a rotating turntable.

The Therac-25 was a successor to earlier devices and adapted both design elements and software from them. However, the Therac-25 improved on its predecessors by eliminating many hardware interlocks that could prevent the beams and objects and such from being put into a dangerous arrangement (such as one that would send too much radiation to the patient, or send it to the wrong part of the patient's body).

Quite literally a "distributed system," the Therac-25 filled a room and was operated from a console outside the room (see Figure 3-1).

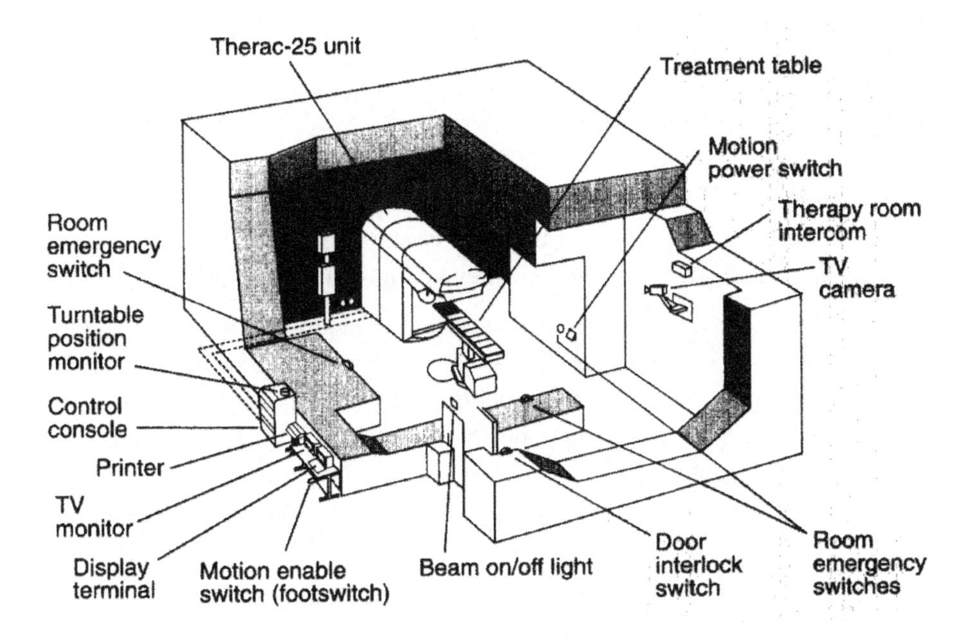

Figure 3-1. Typical Therac-25 setup, from [9]. (Used with permission.)

THE SAD STORY

Unfortunately, the Therac-25 is remembered for a series of incidents in the mid-1980s, when things started going mysteriously wrong. The cases followed a similar pattern: the patient would complain of unusual or unpleasant physical experiences during treatment, such as tingling or burning. In many of the incidents, the operator needed to work with the keyboard to handle some kind of error message.

Outcomes were grim. In Georgia, the patient lost a breast; in Ontario, the patient then required a hip replacement; in Texas and Washington, three patients died. What was happening was that the device was occasionally giving patients massively incorrect radiation doses—Leveson and Turner [8] estimate two orders of magnitude higher than intended. (Subsequent investigation also revealed that many unreported incidents of underdosing may also have been taking place.)

Unfortunately, this behavior was impossible in the mental models of the machine operators, of the system designers, and even of some doctors treating the patients (who suspected radiation burns but could not figure out the causes). This apparent impossibility hampered subsequent efforts to unravel the fact that the machine was doing this—and to discover the series of bugs that were causing these dangerous malfunctions.

The primary bugs were instances of some of the patterns described in "Bug Background" on page 52:

- At least one race condition existed in the operator I/O and machine configuration code—with the result that if exactly the wrong sequence of events happened, the machine would enter a dangerous configuration.
- A critical variable used to indicate whether one of the key components of the beam positioning needed to be tested was implemented as an 8-bit integer and modified by incrementing. As a result, integer overflow could occur, and a nonzero value (256) was represented as a zero—causing the machine to conclude it could skip the test. The Washington death apparently resulted from this overflow occurring at exactly the wrong time.

The fact that exactly the wrong sequence of things had to happen to trigger these errors further complicated detection and diagnosis—repeating the sequence of events reported in an accident would not necessarily trigger the error again.

TODAY

The physical consequences of cyber problems in the Therac-25 provide grist for some software engineering classes, stressing the importance of principled design and testing for such software systems. (The Leveson-Turner paper [8] provides the seminal and exhaustive treatment of the topic.) Unfortunately, even when the focus is reduced from all of smart health to just computerized radiation therapy, cyber issues have still caused problems.

In Evanston, Illinois, in 2009, a linear accelerator made by Varian was used to provide radiation treatment to patients. As with the Therac-25, physical alignment of components was critical to ensuring the beam hit the intended target, rather than healthy tissue. However, the treatment system also required information flow across several computers. At Evanston, "medical personnel had to load patient information onto a USB flash drive and walk it from one computer to another" [3] and had to engage in additional cyber workarounds to get the machine to work with particular focusing components. It's hypothesized that it was because information was lost and changed during translation that some patients ended up having healthy brain tissue irradiated—leaving one 50-year-old woman comatose. Smartening healthcare with IT created more channels where things could go wrong.

At the Philadelphia Veterans Affairs Hospital, clinicians treated patients for prostate cancer with *brachytherapy*, implanting radioactive "seeds" in the cancerous prostates and using a computerized device to make sure the seeds ended up where they were supposed to. However, "for a year, starting in November 2006, the computer workstation with the software used to calculate the post-implant dosages was unplugged from the hospital's network" [11].

As subsequent news coverage and lawsuits attest (e.g., [2]), this lack of closing the loop led to seeds being placed in the wrong places (and sometimes the wrong organs, such as bladders or rectums). Patient cancers received radiation doses too small to be effective, while healthy organs received damaging doses, with serious consequences. Again, more IT created more failure modes.

PAST AND FUTURE

What happened when the smart health future was here before?

Layering IT on top of radiation therapy led to rather bad consequences. The way we built software led to bugs; the way we tied it to physical reality led to the potential of bad outcomes; the way we dispensed with hardware double-checks on the software behavior made that potential a reality. As this future arrives again, we should learn from this past.

Smart Grid

Another IoT application domain receiving much attention is the *smart grid*, layering internetworked computing devices everywhere throughout the massive cyber-physical system that is the electric power grid (Figure 3-2). Some speakers limit the term to just the small portion of the power grid that lies within a consumer's home, but colleagues of mine who are long-time veterans of power engineering actually object to either usage. To them, the electric grid has been "smart" for many decades; the relatively recent push to add even more computational devices is only making it "smarter." Again, to some extent, this future has been here before.

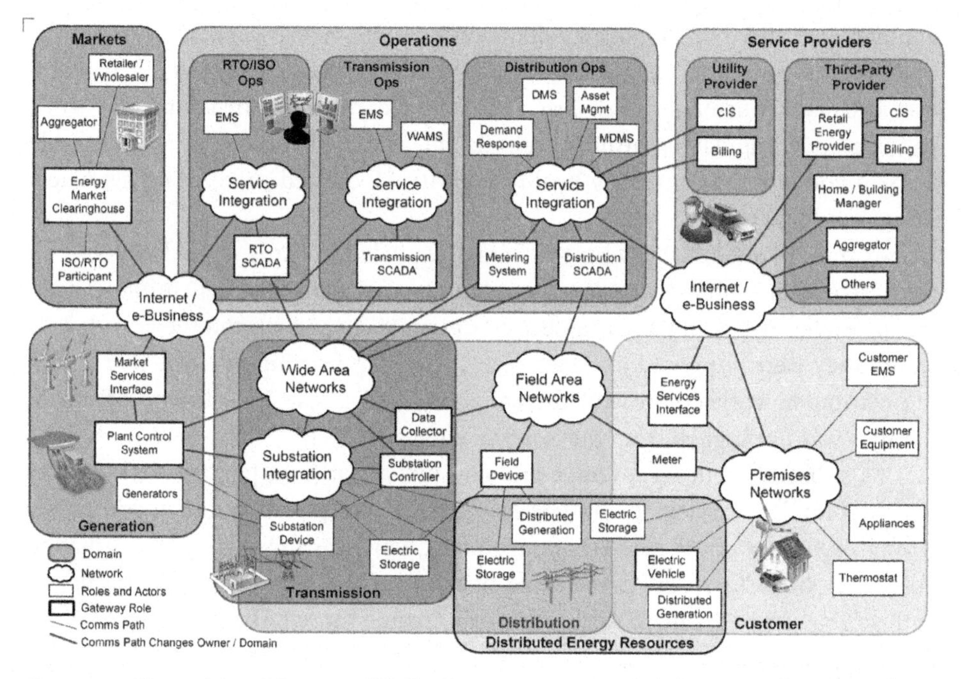

Figure 3-2. The envisioned "smart grid" distributes computational devices everywhere throughout the electricity generation, transmission, and distribution processes. (Source: NIST Framework and Roadmap for Smart Grid Interoperability Standards, Release 3.0.)

THE BALANCING ACT

Electric energy cannot be stored efficiently in large quantities. As a consequence, the giant distributed system of generation, consumption, and routing in between must be kept in near real-time balance—a fact which has long necessitated the use of a layer of cyber communication and control. Making this even more complicated is the subtle interaction between the elements themselves, and between the elements and the greater environment. For example, increased consumption of electricity for air conditioning on a hot day may cause the transmission line carrying this power to heat up and sag too close to trees, and thus require dynamic reconfiguration to send some of this power via an alternate route.

Finally, in the US and many other places, there is no longer a monolithic "Power Company." As a consequence of various economic and government forces, the power grid in fact comprises many players responsible for:

- Generating electricity
- Transmitting electricity (i.e., "long-haul")
- Distributing electricity to consumers
- Buying and selling futures
- Coordinating regional and national-level power

These players may compete in the same marketplaces while trying to cooperate to keep power flowing—a situation some term "coopetition."

One might imagine the "power grid" as a giant network of pipes moving water around. Although perhaps natural, this metaphor can be misleading. One can and does store water in tanks and reservoirs. Furthermore, the movement of an incompressible fluid in a pipe is far more straightforward to mathematically analyze than electric power flow, which requires complex analysis.[2] As a consequence, to determine the state of the grid, we cannot just measure flow in a few places; rather, we need to first measure various aspects of phase angle and such, and then use computers to solve rather hairy differential equations.

The principal motivations for moving from the current grid—already equipped with cyber sensing and control—to the smart grid is to make this balancing problem easier by giving the system more places to measure and compute and switch and tweak.

LIGHTS OUT IN 2003

Interconnection of power across continents and layering cyber sensing and control on top of it can add much stability to these systems, as more opportunities for generation, transmission, and consumption can make it easier to keep the whole system balanced.

2 "Complex" in the sense of $a + bi$, where $i = \sqrt{-1}$.

But unfortunately, cyber failures can translate to physical failures, and the interconnection can cause physical failures to cascade.

A dramatic example of such failure occurred in the northeastern US and southeastern Canada in August 2003 (Figure 3-3). On August 14, a hot day with much power demand due to air conditioning, a power transmission line in Ohio touched a tree and automatically shut off. Ordinarily, grid operators, using standard cyber measurement and control, would have taken appropriate action and kept the system balanced and the lights on. However, for almost two hours that day, the alarm that should have alerted the local control room operators about this line failure itself failed. The humans in the loop, trusted to take correct action based on the information their systems gave them, were subverted because this information did not match reality. Cascading failure followed—along with a flurry of communication as operators throughout the region tried to assess, diagnose, and fix the problem. The end result was an international blackout that, according to the US Department of Energy, affected over 50 million people and cost in excess of $4 billion [10]. According to a later study by epidemiologists [5], the blackout also led to over 100 deaths—all from a tree and a failed alarm.

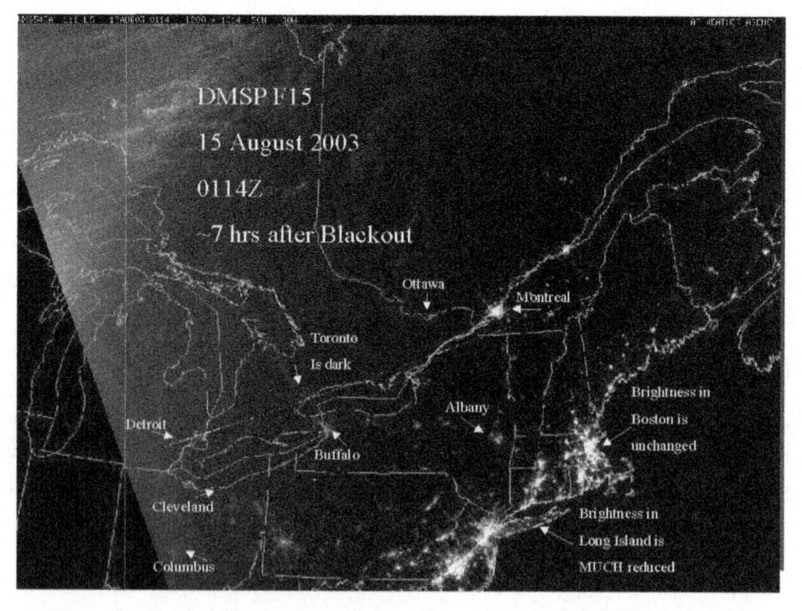

Figure 3-3. The 2003 incident shut down power in the eastern regions of the US and Canada, with the exception of the seaboard. (Source: Air Force Weather Agency.)

ROOT CAUSES

During and after the blackout, much speculation occurred as to its causes. As postmortem analysis unraveled the cascade and identified the failed alarms in the Ohio control room, speculation then focused on why these computers failed to work. Had they been inadvertently infected with run-of-the-mill malware? Or had there been a malicious, targeted cyber attack?

GE Energy, the developer of the software, analyzed the million lines of C and C++ code and eventually discovered a race condition (recall "Bug Background" on page 52) that, under exactly the wrong set of conditions, would cause the alarms to fail. Poulsen [15] quotes GE's Mike Unum:

> *There was a couple of processes that were in contention for a common data structure, and through a software coding error in one of the application processes, they were both able to get write access to a data structure at the same time...and that corruption led to the alarm event application getting into an infinite loop and spinning.*

As noted earlier, concurrent access to shared data structures is a common source of race conditions. As also noted, exhaustive testing may not be sufficiently exhaustive to generate exactly the right wrong scenario. Poulsen observes that "the bug had a window of opportunity measured in milliseconds" and further quotes Unum:

> *We had in excess of three million online operational hours in which nothing had ever exercised that bug. I'm not sure that more testing would have revealed it.*

TODAY

In the tales of radiation therapy recounted earlier, the sad history was followed by a sad coda of more cyber bugs causing more of the same physical consequences.

With this aspect of the smart grid, such a sad coda does not exist (yet). As enticing as it might be for security analysts to identify in the grid what one of my students termed a "Byzantine butterfly" (a tiny but pathological cyber action causing a widespread negative outcome), the closest analysts have come is a Byzantine horde—that is to say, exactly the wrong thing happening in exactly the wrong nine different places at the same time [16].

PAST AND FUTURE

What happened when the future smart grid was here before?

Even in apparently well-engineered and well-tested software, a race condition nonetheless was hiding—and when the right (that is, wrong) circumstances triggered it, it led to cascading failure. As this future arrives again, we should learn from this past: continue to test, but use synchronization more aggressively to prevent race conditions in the first place, and realize that just because sensing and control is cyber-enhanced does not mean it is sufficient. Even with regard to the 2003 blackout, the IEEE Power Engineering Society recommended improving "understanding of the system" and "situational awareness" [1].

That the coming "smarter grid" will greatly expand the number of computational devices, and hence the attack surface—and the tendency of computing devices to use common software components with common vulnerabilities—may also make it easier for Byzantine butterflies to turn into Byzantine hordes.

Smart Vehicles

Another IoT application domain receiving much attention (e.g., [21]) is the *autonomous vehicle*. Some of this attention may be due to the universality of driving a car in many Western societies—self-driving cars may fundamentally transform and maybe even subvert our cultural identity! Another possible cause is the potential positive impact, given the number of fatalities due to traffic accidents, the amount of time Americans waste while commuting in cars, the number of driven miles squandered on seeking parking, and the amount of real estate wasted on parking spaces.

But for whatever reason, this is an IoT area where it's not hard to hear the Cassandras. Surrendering to a computer the basic human right of controlling this transport vehicle is a recipe for disaster—a computing machine (from this point of view) can never duplicate the perception and judgment of a human, so we will have accidents and carnage.

However, this future too has been here before.

THE DAWN OF FLY-BY-WIRE

Let's go back again to the 1980s and think about airplanes.

In the then-current way of doing things, humans in the cockpit used their muscles to move levers and devices that connected via wires and cables to the ailerons and flaps that controlled the flight of the airplane and ensured the passengers safely reached their destination. However, an emerging idea called *fly-by-*

wire threatened this comfortable paradigm. With fly-by-wire, rather than pulling a lever that directly controls the flap, the human pilot would enter some kind of electronic input into a computer, which in turn would actuate some servos that made the flap move.

Placing the computer between the human and the vehicle's moving parts introduces opportunities for the computer to do more than just relay commands. Instead, the computer can actually filter them and participate in the decisions, using its knowledge of the other devices and of the plane's current operating parameters.

To traditionalists, this intrusion sounds dangerous—the computer might not let the pilot do what he or she needs to do to rescue the plane. But by another way of thinking, it could improve safety. For example, if the pilot requests a flap to move at a speed that would be too dangerous, the computer can slow down the transition. Waldrop [20] termed this "envelope protection":

> *Suppose that you are flying along, says Airbus' Guenzel, and you suddenly find yourself staring at a Cessna that has wandered into your airspace. So you swerve. Now, in a standard airliner, you would probably hold back from maneuvering as hard as you could for fear of tumbling out of control, or worse. "You would have to sneak up on it [2.5 G]," he says, "And when you got there you wouldn't be able to tell, because very few commercial pilots have ever flown 2.5 G." But in the A320, he says, you wouldn't have to hesitate: you could just slam the controller all the way to the side and instantly get out of there as fast as the plane will take you. In short, goes the argument, envelope protection doesn't constrain the pilot. It liberates the pilot from uncertainty—and thus enhances safety. As [747 pilot Kenneth] Waldrip puts it, "it's reassuring to know that I can't pull back so hard that the wings fall off."*

In 2017, when cars whose computers assist the driver are already on the road, one hears the same arguments, on both sides.

FEAR OF THE A320

The insertion of computerized interference (or assistance) between the human pilot and the vehicle came to a head when, in the late 1980s, Airbus (a European consortium) introduced the *A320*, a commercial passenger jet that pushed the envelope by being entirely fly-by-wire.

This development led to what sounded like extreme claims on both sides of the spectrum. Airbus claimed that the A320 "would be the safest passenger aircraft ever" (see [7]), in part due to the computer preventing the pilots from putting the aircraft into dangerous operational states. Naysayers worried about computer failures and more insidious errors, such as preventing the human pilots from doing what they knew to be the right thing. A professor of computational mathematics at Liverpool University even took steps to try to stop the A320 in the UK on safety grounds.

(Another factor stoking controversy may have been that the European-backed Airbus was challenging a market previously dominated by American-based Boeing.)

The naysayers seemed to be vindicated in June 1988 when an A320 on a demonstration flight crashed at Mulhouse–Habsheim airport in France, killing three passengers. Although the computer risk community speculated the problem lay somehow with fly-by-wire, the subsequent inquiry vindicated the aircraft and blamed the pilots. Indeed, analysis suggests the software itself is well engineered and robust.

WHAT HAPPENED NEXT

In 1990, another A320 crashed in India and killed 97, engendering more speculation that was squelched again by attribution of the crash to pilot error (ironically, reports indicated that the pilot was attempting to land the plane manually). Then in 1992, an A320 crashed in France and killed 87. This crash was also blamed on pilot error—"cowboy-like behaviour"—but pilots objected [22]:

> *"Each time (one crashes) the crew is blamed whereas the responsibility is really shared in the hiatus between man and machine," said Romain Kroes, secretary-general of the SPAC civil aviation pilots' union.*

In June 2009, an A330 (a later fly-by-wire jet from Airbus) crashed in the Atlantic Ocean, killing all 228 people aboard. Again, speculation of bad computer behavior was rife (e.g., [6]).

The final crash report told a more subtle story, though: the jet's pitot tubes (which measure airspeed) had apparently failed due to ice crystals; confused, the fly-by-wire computer handed control back to the pilots, who then engaged in behavior that brought the plane down. (Episode 642 of NPR's *Planet Money* even went so far as to speculate that if the pilots had done nothing at all, the computer would have taken over again and the crash would not have happened.) There is an echo here of the 2003 blackout—it's hard both for computers and for grid operators to take correct action when they receive input that does not match reality.

PAST AND FUTURE

What happened when the smart vehicle future was here before?

As with self-driving cars now, Cassandras predicted that inserting computers between pilots and vehicles would lead to disaster. The predicted widespread disasters did not happen; indeed, the software engineering in the Airbus jets is held up as an example of what *to* do.

However, the incidents that did occur show that, even without software failure, layering IT into a human-physical system changes how humans work with these systems. The 2009 A330 crash foreshadowed what many identify as the real fundamental problem with autonomously controlled vehicles: not the machine's autonomy but the transfer between machine and humans.

The challenge of interface design—how human operators of equipment, ranging from things as simple as doors and showers to things as complicated as operating system security policies and airplanes—has received much attention over the years. Norman's *Design of Everyday Things* [14] is a great window into this world; Chapter 18 in my security textbook [17] applies this framework to computer security issues. As long-time readers of Peter Neumann's *Risks Digest* know, bad interface design has often been considered as a factor in accidents in old-fashioned nonautonomous vehicles—for example, the death of John Denver when piloting a small aircraft has been attributed to the bizarre placement and orientation of the control for which of the plane's two fuel tanks should be used [18]:

> *The builder of the aircraft, however, elected to place the valve back behind the pilot's left shoulder.... The only way to switch tanks was to let go of the controls, twist your head to the left to look behind you, reach over your left shoulder with your right hand, find the valve, and turn it.... It was difficult to do this without bracing yourself with your right foot—by pressing the right rudder pedal all the way to the floor....*
>
> *He also rotated it in such a way that turning the valve to the right turned on the left fuel tank. This ensured that a pilot unfamiliar with the aircraft, upon hearing the engine begin missing and spotting in his mirror*

> *that the left fuel tank was empty, would attempt to rotate the fuel valve to the right, away from the full tank, guaranteeing his destruction.*

Making vehicles smart brings a new dimension to this human/machine boundary—not just *how* to do the controlling, but *who* is doing the controlling. Chapter 10 will revisit this topic.

TODAY

As Chapter 1 mentioned, summer 2016 brought news of perhaps the first fatality due to a self-driving car: the May 7 death of Joshua Brown in his Tesla in Williston, Florida. Reports sketch a complicated situation—apparently, Brown was speeding and watching a DVD [9]:

> *[The truck driver] said the Harry Potter movie "was still playing when he died and snapped a telephone pole a quarter mile down the road".*

What appeared to be driver carelessness was exacerbated by the Tesla's computer finding itself in a confusing physical scenario [4]:

> *Tesla says, "the high, white side of the box truck ... combined with a radar signature that would have looked very similar to an overhead sign, caused automatic braking not to fire." This is in line with a tweet from Tesla CEO Elon Musk yesterday, who suggested that the system believed the truck was an overhead sign that the car could pass beneath without incident.*

Subsequent reports suggest Tesla has worked on algorithm updates to prevent this kind of scenario from reoccurring.

Indeed, I am tempted to put on my professor hat and observe that real world is often messier than that of simple models. It would be easier to blame either the self-driving car (watch out, a dangerous future!) or the foolish human (the future is the best of all possible worlds!), but—according to initial reports—both bore some responsibility.

Some initial comments from Tesla also reflect the messy nature of reality [9]:

> *Elon Musk, the CEO of Tesla, tweeted his condolences regarding the "tragic loss", but the company's statement deflected blame for the crash. His 537-word statement noted that this was Tesla's first known autopilot death in roughly 130m miles driven by customers.*

"Among all vehicles in the US, there is a fatality every 94 million miles," the statement said.

Self-driving cars may reduce the total number of traffic fatalities, but they may also change who gets killed—which is good overall, unless one happens to be one of the ones who gets killed. As Joshua Brown might have rephrased John Donne, "Any man's death diminishes me—especially mine."

Not Repeating Past Mistakes

What did these past experiences tell us about the IoT?

Sloppy software engineering enhances the chance of software failures, and integration with the physical environment means these failures have physical consequences. However, even well-engineered software can still hide bugs. Even without failures, the cyber-physical connection adds a new surface to the system, and the solution to problems in cyber-physical control might sometimes be to add more cyber.

Have these lessons been taken to heart? In 2013, Toyota agreed to pay a $1.2 billion fine due to safety issues regarding sudden acceleration in its cars [19]. Technical analysis of this phenomenon identified numerous sloppy coding practices: race conditions, failure to detect bad combinations, memory corruption, and the like [13]. With Toyota vehicles—and all the other smart IoT devices being rushed to market—might we have more Therac-25s lurking?

Chapter 4 will consider software engineering factors that lead to such problems, Chapter 7 will consider the economic factors, and Chapter 8 will consider public policy factors.

Works Cited

1. G. Anderson and others, "Causes of the 2003 major grid blackouts in North America and Europe, and recommended means to improve system dynamic performance," *IEEE Transactions on Power Systems*, October 31, 2005.

2. W. Bogdanich, "At V.A. hospital, a rogue cancer unit," *The New York Times*, June 20, 2009.

3. W. Bogdanich and K. Rebelo, "A pinpoint beam strays invisibly, harming instead of healing," *The New York Times*, December 28, 2010.

4. J. Golson, "Tesla and Mobileye disagree on lack of emergency braking in deadly Autopilot crash," *The Verge*, July 1, 2016.

5. K. Grens, "Spike in deaths blamed on 2003 New York blackout," *Reuters Health News*, January 27, 2012.

6. J. T. Iverson, "Could a computer glitch have brought down Air France 447?," *Time*, June 5, 2009.

7. N. Leveson, "Airbus fly-by-wire controversy," *The Risks Digest*, February 23, 1988.

8. N. Leveson and C. Turner, "An investigation of the Therac-25 accidents," *IEEE Computer*, July 1993.

9. S. Levin and N. Woolf, "Tesla driver killed while using Autopilot was watching Harry Potter, witness says," *The Guardian*, July 1, 2016.

10. B. Liscouski and others, "Final Report on the August 14, 2003 Blackout in the United States and Canada." U.S.-Canada Power System Outage Task Force, April 2004.

11. M. McCullough and J. Goldstein, "Unplugged computer cited in Phila. VA medical errors," *Philadelphia Inquirer*, July 19, 2009.

12. M. Miller, *The Internet of Things*. Que, 2015.

13. P. Mundkur, "Critical embedded software bugs responsible in Toyota unintended acceleration case," *The Risks Digest*, October 26, 2013.

14. D. Norman, *The Design of Everyday Things*. Basic Books, 2002.

15. K. Poulsen, "Tracking the blackout bug," *SecurityFocus*, April 7, 2004.

16. R. Smith, "U.S. risks national blackout from small-scale attack," *The Wall Street Journal*, March 12, 2014.

17. S. Smith and J. Marchesini, *The Craft of System Security*. Addison-Wesley, 2008.

18. B. Tognazzini, "When interfaces kill: John Denver," *Ask Tog*, June 1999.

19. E. Tucker and T. Krisher, "Toyota to pay $1.2 billion fine for misleading consumers on sudden acceleration," *The Dallas Morning News*, March 19, 2014.

20. M. Waldrop, "Flying the electric skies," *Science*, June 30, 1989.

21. P. Wayner, *Future Ride v2*. CreateSpace, 2015.

22. M. Wrong, "Latest crash heightens controversy over Airbus A3320," *Reuters*, January 21, 1992.

4

Overcoming Design Patterns for Insecurity

If we build the IoT like we did the IoC, we're going to have security trouble.

"Security," as I like to define it, is the extent to which a system remains in a correct state, despite the efforts of a malicious adversary, perhaps in conspiracy with an uncooperative universe. Typically, the root cause for a successful attack lies in a design or implementation mistake in the system: some error or feature that enables the adversary to come in. These tend to fall into general patterns—and since IT systems have existed for a while now, it's often hard to come up with completely new ways to do things wrong.

When feeling a bit arrogant (or frustrated), security specialists (such as me) are tempted to label these mistakes as "blunders." These are not subtle and complex failure cases, but grievous oversights that are obvious—or at least they appear obvious to security specialists looking at them in hindsight. The problem, of course, is that these design decisions need to be made beforehand, and by people who are wrestling with many priorities (and do not realize that *my* priority is the most important).

This chapter surveys and explains some of the principal categories of IoC blunders ("anti-patterns") that have affected (or are likely to affect) the IoT:

- Doing too much
- Coding blunders
- Authentication and authorization blunders
- Cryptography blunders

However, the point here is not that hope is lost, but rather that sensibly moving forward will require an awareness of these patterns and some creative thinking to address them.

Anti-Pattern: Doing Too Much

Helpful employees do more than what's expected of them; helpful people do their best to understand what was meant by faulty and flawed communication. However, a standard source of security trouble is when system interfaces provide more functionality than they are supposed to, perhaps by accepting and acting on incorrectly formatted input. Such extra functionality can allow an adversary literally to take control of a system by whispering the right bizarre magic words; even helpfully "correcting" incorrect input can lead to trouble when two systems do it differently.

INSTANCE: FAILURE OF INPUT VALIDATION

In theory, an interface expects inputs following some specific formats; for example, a name of 20 characters here, a number in this range there. In practice, interfaces often accept much more but still implicitly assume the inputs are formatted correctly. As a consequence, in the IoC, perhaps the most common way to attack a system is to give it deviously *crafted input* that does not fit the rules of what the programmer intended as valid input but which the system accepts nonetheless and which tricks the system into carrying out incorrect behavior.

The classic instantiation of this pattern is the *buffer overflow* attack (e.g., [1], or Section 6.1 in [49]). Here, the victim system expects a character string from the user and copies it into a fixed-length buffer in the stack frame—but never checks that the string actually fits within the buffer. Consequently, an adversary can provide a very long string crafted to include some executable code and to overlay the "return address" field on the stack frame with the address of this code. When the system's current function returns, it then begins executing code the adversary injected.

This family of attacks has a long and storied history of variations, countermeasures, countermeasures to the countermeasures, and so on. Unfortunately, but perhaps not unexpectedly, this pattern is already manifesting itself in the IoT. Crafted input attacks have emerged targeting Android [5] devices. IoT thermostats have allowed unprivileged remote adversaries to inject code via network packets that caused a buffer overflow [12]. Home cable modems have allowed unprivileged remote adversaries to inject code via device web requests [16]. In spring 2016, a researcher discovered an input validation failure in firmware used in many dozens of models of CCTV equipment that enabled an adversary to run privileged code by sending a deviously constructed URL to the device's web server interface [13].

In recent years, researchers have even begun to address the mathematical foundations of this type of attack as a problem recognizing the "formal language" of valid inputs [46]. My own lab has produced tools that have found holes in power grid control systems via *fuzzing* (that is, automatically modifying otherwise correct input) [47]. Both of these fronts may be helpful in mitigating the pattern in the IoT.

INSTANCE: EXCESS POWER

Besides accepting too much, another pattern we see emerging in the IoT is *doing* too much. Systems may have more power than they need; and as with the principle of least privilege (taught in security textbooks), having excess power can lead to excess damage should the system be compromised.

One example here that my lab found (discussed further in "Anti-Pattern: Authentication Blunders" on page 75) involved a commercial set-top box with a full Linux kernel inside. The box offered some small obstacles to penetration and, once penetrated, provided obstacles to prevent us from adding malicious software to the box itself. However, the box inexplicably had support for the NFS remote filesystem, so we could install our malware simply by having the box remotely mount a remote disk. There was no sensible reason why the box needed NFS support; we hypothesize that it was there simply because it was part of the Linux bundle. Using established components also opens a system up to established attacks—as a colleague's hospital learned when its IT went down for a week due to a virus infecting the Windows 95 installation hiding inside a radiology machine.

Using a tried-and-true, off-the-shelf component rather than building a new one is good engineering practice, so it often makes sense to choose some standard bundle. However, in cases like these, it opens the door to too much. (An IoT security colleague laments that every IoT device will eventually run full Linux "because it can.") Moving forward, we may need to rethink the use of tried-and-true IoT components.

INSTANCE: DIFFERENTIAL PARSING

The system (attempting to be helpful) can, in fact, do too much when it receives malformed input. For example, multiple different systems, each allegedly speaking the same input language, may take different actions when presented with deviously crafted input. This behavior enables *differential parsing*, crafting input that only selected listeners will hear, because the others will discard it as malformed (e.g., [34]).

Such attacks have already been demonstrated for wireless stacks used in IoT devices [28, 33].

Key to this research result were tools the researchers built to probe stack interfaces with crafted input—something otherwise difficult in vertically integrated IoT systems. In the IoT, it will be important to think not just of parsing flaws, but also of ensuring the interfaces can be tested before it's too late.

Anti-Pattern: Coding Blunders

As any programming student knows, it's easy to make mistakes. The software industry is plagued with errors and bugs, and (in the case of larger, recognizable computers) often depends on a process of "penetrate and patch," as mentioned earlier, to keep things going. But this penetrate-and-patch approach poses problems for the IoT (especially if the "things" no longer look like computers needing updates), leading to the risk of the forever-days discussed back in Chapter 1.

The IoT has already suffered from its share of coding blunders seen in the IoC. Machines can get infected; body cameras (for the "Internet of First Responder Things") have shipped with malware installed (presumably unintentionally), ready to infect the systems to which they are connected [10]. Local device action can have global network impact: in one reported case, a smart lightbulb burned out and sent out so many announcements about this fact that it overwhelmed and shut down the smart home's wireless network [32].

Debug code has been inadvertently left in shipped products, too. A SANS researcher discovered his Netatmo Weather Station still contained debug features that would transmit memory contents—including his home WPA password—back to home base, unencrypted [56].

In the IoC, blunders (as well as more subtle flaws) can be fixed via patching. In the IoT, however, we are already seeing zero-days becoming forever-days. In one example, researchers analyzing a code injection flaw in Android discovered 60 percent of a sample of a half million phones were vulnerable—and "27 percent of those devices were found to be 'permanently vulnerable' in that they are too old to receive monthly updates" [4]. And as we saw earlier, in 2015 Trend Micro reported that over six million consumer IoT devices were still vulnerable to a code injection flaw that was patched in 2012 [59]; the patches are just not getting there.

As mentioned in many places in this book, "penetrate and patch" may work for the IoC, but it's not likely to work in the IoT unless we rethink patching—or perhaps rethink our coding and testing practices to reduce the need for patching.

Anti-Pattern: Authentication Blunders

It's important that a system verify who is claiming to send it an update, or change its configuration, or act as its absolute master. The IoT has some aspects that make this particularly critical. The intimate connection of IoT systems to the real world can make the consequences of malfeasance more severe than in the IoC. The distribution of an IoT system throughout the real world can greatly increase the attack surface—places the adversary can touch. Finally, effective authentication can be a question of management; the larger and more disorganized a set of entities is, the harder it can be to manage, and it is hard to think of something larger and more disorganized than the aggregation of banal real-world things.

The IoC manifests many common "design patterns for insecurity" with respect to authentication:

- A service may require *no authentication* whatsoever.
- A service or set of services may use *default* authentication credentials, easily discoverable.
- A service may have a *permanent credential*, never changeable.
- A service may fail to allow for *revocation* of a credential or privilege.
- A service may fail to allow for *delegation* of a privilege to another legitimate party (thus engendering workarounds).

These patterns are already emerging in the IoT. ("Anti-Pattern: Cryptography Blunders" on page 81 will consider more specific cryptographic flaws.)

INSTANCE: NO AUTHENTICATION

One of the most extreme ways a system can fail at authentication is to overlook it completely and provide a given service to anyone who requests it. Such a design can arise when the designers either did not think about security at all, or had a security model that theoretically disallowed the adversary from even reaching this service.

A timely IoT example of this is the controller area network (CAN) bus central to the IT system in a modern car. A component on this bus listens to see if it is being spoken to—but it has no way of verifying who was doing the speaking; rather, it implicitly assumes that anyone speaking to it has the right to do so. For example, it's the engine control unit (ECU) that is supposed to tell the brakes to engage, but anything on the CAN bus can actually do this actually do this.

This flaw lays at the heart of the attacks Stefan Savage's group published in academic venues in 2010 and 2011—including injecting malware into the car's CD player, which then spoofed commands on the CAN bus [9]. It was also the weakness exploited in the highly publicized shutting down of a *Wired* reporter's car (via the cellphone connection) while it was being driven on the highway in 2015 [29]. Researchers subsequently demonstrated attacking the CAN bus via digital radio [57]. There are even firsthand reports of a commercial automobile with a CAN bus connection in the taillight—so even if the car is locked, an adversary can jack in, unlock it, start it, and have all sorts of fun.

Chapter 1 lamented the difficulty of updating software on smart devices. Unfortunately, the opposite can also be a problem: some devices permit updating with no authentication whatsoever—if someone claims to have an update, the device happily reprograms itself. In the attack on the Ukrainian power grid in late 2015 [19], this flaw was used to turn many of the field devices into inert bricks that were beyond field repair, since the updated "software" no longer provided an update feature. This flaw was also discovered in Progressive Insurance's Snapshot devices, which consumers plug into the diagnostic port in their cars to enable Progressive to measure their driving behavior and charge them lower premiums if the measured behavior implies lower risk [23]. An adversary who gets into this device can then get into the car's internal network and launch an attack as just described. Direct physical access to the Snapshot would permit the adversary to inject a malicious upgrade. Compromise of the modem connecting the Snapshot to Progressive's back-end would also enable the adversary to do this—and (as [23] notes) similar devices have been compromised. In 2016, CEO of EyeOs J. C. Norte wrote of discovering a different way to remotely reach into the open CAN network: via internet-connected *Telematics Gateway Units* (used at automotive repair shops) which themselves offered services worldwide with no authentication [44].

For a higher-cost example of consequences, in 2016 the Japanese space agency lost a $286 million satellite, apparently due to a bad (although, in this case, not malicious) software update [41].

A variation on the "no authentication" pattern is "insufficient authentication." An interesting example of this in the IoC occurred several years ago [50] when an online third party handling business school applications permitted applicants to learn admissions decisions early by logging in as themselves but then editing the URL to request services for which they were not yet authorized. This pattern has also surfaced in the IoT. For example, CoreLabs discovered that a projector commonly used in smart classrooms requires authentication to go from the *index.html* page to the interior *main.html* page, but skips authentication if the user opts to go to directly to *main.html* [7].

In a summer course I taught on risks of the IoT, students used the Shodan search engine [39] to scan for IoT-connected devices on the internet at our own university, and found printers, smart classroom devices, and videoconference equipment requiring no passwords. In the ethical hacker space (e.g., see the work of Dan Tentlar or Paul McMillan), researchers have discovered tens of thousands of control interfaces for IoT-style cyber-physical systems—with much scarier misuse consequences—on the open internet, with no authentication required, via the Virtual Network Computing (VNC) protocol.

In another variation, Rapid7 discovered a family of internet-connected industrial control systems that required authentication credentials over an SSH connection, but would accept *any* credentials [55]. Similarly, the "Hello Barbie" internet-connected doll would try to authenticate its network via SSID name, but would accept anything with the name "Barbie" [43].

INSTANCE: DEFAULT CREDENTIALS

Designers who do in fact include authentication in their systems are then faced with a challenge: what should the system do when it first comes out of the box? A common approach is to ship systems preinstalled with a default password. Unfortunately, these default passwords are usually well known (students find them easily with web searches); and given the natural human inclination to choose the path of least resistance, default passwords often are never changed.

This problem is already surfacing in the IoT. Trend Micro reported finding a backdoor account with a hardcoded, common password in over two million routers [35]. Researchers at EURECOM reported finding "hard-coded web-login admin credentials" in over 100,000 internet-facing IoT devices [18]. My own lab was involved in discovering a commercial set-top box with a default password for root [2]. My students also reported finding a programmable logic controller—the same kind of beast exploited in the Stuxnet attack—used in industrial control, on the internet with a default password. Default SSH backdoors have even been discovered in security devices from Cisco and Fortiguard [20, 17].

For perhaps a more tangibly creepy manifestation of the consequences of default credentials in real life, it can be enlightening to consider network-connected "security" cameras intended to let homeowners and such monitor their households and families from over the network—but which, due to the use of default passwords, let *anyone* do that. A "Ms. Smith" in *NetworkWorld* (no relation, as far as I know) writes of a collection of over 73,000 such devices that she discovered were peering into backyards and living rooms and children's bedrooms. Many similar images can be found with some poking around [48].

For a scarier application domain, *ZDNet* reports [58]:

> *An application suite designed to help clinical teams manage patients ahead of surgical operations includes a hidden username and password, which...could allow an attacker to "backdoor" the app to read or change sensitive information on patients, who are about to or have just recently been in surgery.*

Ars Technica earlier warned of a similar permanent, built-in account in RuggedCom's Rugged Operating System, ironically intended for high-assurance devices in industrial control systems [24].

And as I was writing these words, a news report came in from Canada [22]:

> *Two 14-year-old high school students managed to hack into a Bank of Montreal ATM at a super market during their lunch break using an operator's manual they found online.*
>
> *When they brought up the administrator mode screen it asked for a password, to which the teens used the factory default password. To their surprise it worked.*

INSTANCE: PERMANENT CREDENTIALS

An authentication credential is an electronic embodiment of the right of some entity to invoke some service. This expression involves three relations:

- A binding between entity and right
- A binding between entity and credential
- A binding between credential and right

Trouble can emerge when not all of these bindings are permanent.

One pattern is when a right is permanently bound to a credential—it is embedded in an IoT device itself and can never be changed. For example, in the set-top box mentioned previously, one could change the root password, but the password would then return to the default after rebooting.

A default password gives the entire world the right to become root. Hardcoding the password makes it impossible to deny that right to adversaries—which also applies in the Trend Micro and EURECOM results mentioned in the previous section. ComfortLink thermostats had hardcoded credentials preinstalled [12]. Penetration specialist Billy Rios found the same thing in an engagement at the Mayo Clinic [45].

Another pattern is when an entity may indeed possess the right to use a service at some T_0, but then loses that right after some time $T_1 > T_0$. If this entity still knows the required credential, then one needs to be able *revoke* the binding of the credential to the service—otherwise, one permits unauthorized access. However, in the rush to make things work now, it can be easy to overlook revocation. For an amusing example, *InfoWorld* reported on a web-controllable thermostat that received this favorable review on Amazon [42]:

> *Little did I know that my ex had found someone that had a bit more money than I did and decided to make other travel plans. Those plans included her no longer being my wife and finding a new travel partner (Carl, a banker). She took the house, the dog and a good chunk of my 401k, but didn't mess with the wireless access point or the Wi-Fi enabled Honeywell thermostat.*
>
> *Since this past Ohio winter has been so cold I've been messing with the temp while the new love birds are sleeping. Doesn't everyone want to wake up at 7 AM to a 40 degree house? When they are away on their weekend getaways, I crank the heat up to 80 degrees and back down to 40 before they arrive home. I can only imagine what their electricity bills might be. It makes me smile. I know this won't last forever, but I can't help*

> *but smile every time I log in and see that it still works. I also can't wait for warmer weather when I can crank the heat up to 80 degrees while the love birds are sleeping. After all, who doesn't want to wake up to an 80 degree home in the middle of June?*

The article observed that "more than 8,200 of the 8,490 Amazon users who have read the review deemed it 'useful.'"

INSTANCE: NO DELEGATION

Another challenging aspect of authentication is determining who should be granted access in the first place. (Chapter 6 will look at a similar problem: that of effective creation of privacy policy.) One trouble pattern here arises from keeping this policy creation too far out of the hands of the end users themselves. When end users cannot easily configure a system to permit services they believe should be permitted (or when such configuration is not even possible), then users will often circumvent protections, breaking the security system [53].

For one IoT example of this inflexibility, consider the power grid. One vendor of equipment in this space had a marketing slide showing the default user ID and password on its equipment, and the default user IDs and passwords on equipment from its competitors. The point of this slide was to stress the security of this vendor, since its default password was so much harder to guess than the defaults for the competitors. Security specialists react to this slide with laughter —how can a system with a default and published password be secure? Power specialists react differently; this application domain requires that, in emergencies, repair crews (perhaps borrowed from another region) need to be able to quickly access a device, so a system that does not allow easy and dynamic granting of access does not work. Default passwords provide this feature, which they regard as critical.

Smart medicine deployments reveal similar problems with policy inflexibility [52]. One smart medication administration system ensured that a nurse could only give a patient exactly the medicine and dosage prescribed—a policy that did not account for the reality that there might not be any 20 mg tablets left, but that two 10 mg tablets provide a 20 mg dose. Another enforced that a 10-hour medication regimen must begin exactly when the issuing doctor said, even if the patient did not arrive from dialysis until an hour later.

INSTANCE: EASY EXPOSURE

Another pattern for insecurity is when the infrastructure supporting authentication itself subverts it. One IoC example is all the websites that require a user ID and password (good), but collect them over a plain-text channel (bad—anyone can listen in!) or through a pop-up basic authentication window (also bad—the user cannot easily tell who offered this window). We see similar things happening in the IoT: the set-top box mentioned earlier only provided unencrypted `telnet` access for root, not `ssh` (the literature notes a few thousand other similar discoveries); wind turbines have plain-text passwords embedded inside them [21]. The EURECOM researchers extracted "several dozens of hard-coded password hashes" and over 35,000 RSA private keys from the devices they surveyed [18]. (For more on hashing, see "Cryptographic Hashing" on page 101.) Other examples abound:

- It is rumored that one enabling aspect of the Ukrainian power attack was the grid's internal use of easily harvestable credentials.
- In 2014, researchers discovered an IoT alarm system that used weak authentication mechanisms—and, furthermore, transmitted them unencrypted [11].
- In 2016, researchers from the University of Michigan published a way [27] to obtain a user's credentials for Samsung's *SmartThings* smart home devices: tricking the user into entering them into a site controlled by the adversary—the IoC web-spoofing pattern taking root in the IoT.

MOVING FORWARD

All these authentication issues are already problems in the IoC; solving them requires thinking carefully about the players and systems and deployment patterns involved, and the authentication and authorization technology to support this. In the IoT, the number of moving parts scales way up; Chapter 5 will consider some resulting challenges.

Anti-Pattern: Cryptography Blunders

Cryptography, the mathematical and computational systems to transform data into formats that make it harder for inappropriate parties to extract useful information, is central to implementing many authentication and authentication-related techniques. ("The Standard Cryptographic Toolkit" on page 95 will give more details on the basics of cryptography for the IoT.)

Cryptography is used in the IoT for the same reasons it is used in the IoC: for parties to protect information (confidentiality and integrity) while also working with it in various ways. An IoT device communicating to a backend server over an open channel would use cryptography to protect the data in transit. IoT systems (including backend servers) storing sensitive data would use cryptography to protect it at rest. An IoT device trying to authenticate the provenance of some data (e.g., did this firmware upgrade really come from the right party?) would use cryptographic signatures or message authentication codes (MACs). IoT entities trying to authenticate to one another without actually giving their authentication secrets away would also use cryptography.

INSTANCE: BAD RANDOMNESS

As Chapter 5 will discuss, the security of cryptography typically requires that a key is known only by the parties who are supposed to have the privilege that the key embodies. If the keys get revealed, the rest of the system breaks (and indeed, a colleague who worked on nation-state cryptanalysis would quip that it was always easier to steal the keys than to break the mathematics).

In using a public key algorithm such as RSA, the first step is for a key-owning entity to generate its keypair. This process requires using unpredictable randomness—for RSA, the entity uses this randomness to generate a pair of prime numbers which an adversary would not be able to predict. However, in a study surveying (at internet scale) internet-facing machines with keypairs for the TLS and SSH protocols, researchers from UC San Diego and the University of Michigan [29] found something disturbing. A large number of machines either shared the same keypairs or had shared one of these prime number factors (and since efficient algorithms are known to calculate the GCD—greatest common divisor—of two large integers, this suffices to break these keypairs).

They concluded that an underlying problem here might be simple embedded IoT devices that used standard key generation software on standard operating systems such as Linux. For random seeds, these standard tools use */dev/urandom*, which gets its entropy from things such as keystroke timing and disk movement and previous memory contents—possibly reasonable on a standard computer, but (as the researchers verified experimentally) highly predictable when these tools are moved to headless IoT systems. The researchers lamented this "boot-time entropy hole." Moving a standard, trusted component from an IoC device to an IoT device turned out to invalidate implicit assumptions.

This academic paper predicted repeated keys in IoT systems. Interestingly, this prediction appears to have come to pass. For example, in 2015, *Network World* reported on Shodan finding populations of more than 100,000 devices sharing common keypairs [37]. A month later, *ITworld* reported on researchers finding 28,000 devices using the same keypair [36]. What's next?

This blunder also connects to the problem mentioned earlier: conventional wisdom says it's better to use a tried-and-true component (e.g., a Linux installation with standard key generation code) than to build something new from scratch; but once more, moving a standard IoC solution to an IoT device created new problems. Researchers have already presented some interesting (and scary) work on how reuse of key-generation algorithms may lead to more subtle weaknesses [54].

INSTANCE: COMMON KEYS

Not just RSA but also symmetric cryptography can have problems when keys are repeated. In 2014, security researchers found an example of this in LIFX internet-connected lightbulbs [25, 8]. The implementation encrypted transmissions using the NIST-standard AES algorithm, but used the same common key in all LIFX bulbs—enabling easy snooping by any adversary. (Admirably, after the researchers informed LIFX, the flaw was fixed.)

INSTANCE: BAD PKI

Working with public key cryptography, a *relying party* might be able to conclude that some remote entity actually knows the private key matching some public key. However, to draw a meaningful conclusion from this information, the relying party usually also has to know something else about that public key, for example, that knowledge of the matching private key means the entity has some particular identity.

Public key infrastructure (PKI) is the term used for all the glue and machinery necessary to establish this additional information. Typically, PKI rests on *certificates*: statements asserting that the owner of some public key has some given properties. These statements are themselves digitally signed by a *certification authority* (CA) who presumably is in a position to know. To draw a conclusion about a keyholder, a relying party needs to obtain a set of certificates chaining from a *root* (whom the party trusts) which satisfies some logical calculus; part of this satisfaction includes verifying that a given certificate has not been *revoked*.

When it comes to PKI, the IoC itself shows many standard trouble patterns:

- The relying party fails to check whether a certificate has been revoked. One reason this happens is the performance overhead of doing this checking.
- The keyholder uses a certificate issued by a bogus CA, such as itself (a *self-signed certificate*—often the default consequence of standard tool installations).
- More subtly, the keyholder information established by the standard PKI may not match what the relying party wants to know. For example, Chris Masone's Ph.D. work [40] explored how standard S/MIME email PKI would not suffice to reproduce how power grid operators authenticated one another over the telephone in the 2003 blackout recovery, because S/MIME did not express what users needed to know. Reality has a more complicated and nuanced ontology than straightforward identity PKI allows for.

We expect revocation and ontology issues to surface in the IoT (see Chapter 5) but have no war stories to share yet—except to observe that Android does not support certificate revocation, and Symantec laments its lack in other IoT systems [3]. We have already seen the bogus CA pattern emerge. As the researchers from UC San Diego and the University of Michigan [30] noted:

> *At least...85,046 TLS hosts (0.66%) served default Apache certificates (sometimes referred to as snake-oil certificates, because they often include the CN* www.snakeoil.com*).*

(When my students went hunting, they found many invisible computers, such as printers, offering self-signed certificates.)

However, the IoT has already demonstrated a new trouble pattern here: failing to check certificates at all. A high-profile example of this pattern emerged in 2015, when researchers discovered that Samsung's smart fridge would open an SSL-protected channel to Google—but without actually checking whether the certificate from the alleged Google server was valid [38]. This flaw permits a "man in the middle" to pretend to be Google and collect the consumer's Gmail credentials and other personal information. Similarly, the *certifigate* bug family in Android featured many ways code was authenticated via a signature and certificate, but where the adversary could subvert the means by which a certificate was accepted as the right one [6]. Similar reports exist for other IoT products (e.g., [15]).

Again, Chapter 5 will consider these trust management issues when we scale from the IoC to the IoT.

INSTANCE: AGING OF CRYPTOGRAPHY AND PROTOCOLS

Looking over the last several decades, one can see many issues where cryptography did not hold up with the passage of time. Keys that were considered long enough to last decades weakened much more quickly. Algorithms such as DES slowly weakened; algorithms such as MD5 or ISO9796-1 weakened dramatically.

The potential long lifetime of IoT devices, coupled with the difficulty of updating their software, suggests a potential new trouble pattern: *devices that last longer than their cryptography*.

A version of this pattern has already emerged in both the IoC and the IoT: protocols designed to be flexibly compatible with various variations end up backward-compatible with variations now considered insecure. In the IoT, the backend servers to which the "Hello Barbie" toys communicate turned out to be using an SSL implementation that could be tricked (via *POODLE*) into using weak key lengths. Researchers have found millions of instantiations of mobile applications vulnerable to the similar *FREAK* attack [14].

Today, one would not trust commodity encryption considered secure in 1990. Will the world of 2045 trust commodity encryption considered secure in 2017? If not, what will we do about the forever-day IoT devices we release in 2017 that are still out there in 2045? Furthermore, problems in a backend server in a data center should easily be fixable by standard IoC patch techniques—but what about IoT toys distributed in homes internationally?

A Better Future

IoT systems will almost certainly suffer from the design patterns for insecurity that have plagued the IoC.

Managing this problem will likely require a mixture of more thorough application of good engineering principles that are already known and development of new techniques and paradigms.

One straightforward approach to this problem is a renewed effort at better security awareness and education for IoT developers—crucial since their products' attack surfaces and physical reach are so substantial.

When it comes to input validation problems, we might try more principled approaches to specifying and recognizing valid input (e.g., [46]). We might also try new kinds of fuzz testing to discover where validation has failed; for instance, for power-grid SCADA, our lab needed to develop a new kind of fuzzing tool that learned quickly from live input, since a corpus of archived canned input acceptable to a client was not available. To mitigate general blunders, we might learn from application domains such as the telephone network (which developed industry-strength formal model checking tools after an unexpected global effect from local action took down the network in the 1980s), or from the design and testing regimen used in high-reliability software such as fly-by-wire aircraft (Chapter 3).

To mitigate forever-days, we might want to consider combinations of:

- Making sure the vulnerabilities can be discovered and patches can be created.
- Making sure patches can be pushed.
- Making sure someone still exists to push the patches.
- Making sure the patches do not introduce worse, unknown bugs.
- Making systems automatically die off, as telomeres enforce in cell biology, if they are not patched (although this feature might have the problem of annoying consumers).

Adding authentication to IoT systems will require a more careful enumeration not only of the policy requirements (recovation? delegation?), but also of the performance requirements. It may very well be that in the case of the CAN bus, as with other specialized control systems (e.g., [51]), timing and data requirements may in fact make textbook security techniques infeasible, and we will need new ways of thinking.

Surveying the recent literature for IoT security design flaws also reveals hints of mitigation techniques. For example, the researcher who discovered the debug exfiltration in the Netatmo Weather Station did so because he had an automatic guard in place looking for his WPA passphrase being transmitted in plain text. Discussions of authentication failures in home routers also include an article noting that the US Federal Trade Commission has taken action against one vendor [26]. Analyses of the weaknesses of smart home devices also can include some good-sense advice on tightening things up [31].

Fixing the future is going to take a combination of big battles and little details, including future-proofing cryptographic authentication (Chapter 5), balancing economic forces (Chapter 7), and crafting public policy and law (Chapter 8) to promote more careful software engineering.

Works Cited

1. Aleph One, "Smashing the stack for fun and profit," *Phrack*, November 8, 1996.
2. K.-H. Baek and others, "Attacking and defending networked embedded devices," in *Proceedings of the 2nd Workshop on Embedded Systems Security*, October 2007.
3. M. Ballano Barcena and C. Wueest, *Insecurity in the Internet of Things*. Symantec, March 12, 2015.
4. D. Bisson, "Attackers can pwn 60% of Android phones using critical flaw," *Graham Cluley*, May 23, 2016.
5. D. Blison, "New Stagefright exploit threatens unpatched Android devices," *Graham Clulely*, March 18, 2016.
6. O. Bobrov and A. Bashan, *Certifi-gate: Front Door Access to Pwning Millions of Android Devices*. CheckPoint, July 18, 2015.
7. C. Brook, "Authentication vulnerabilities identified in projector firmware," *ThreatPost*, April 28, 2015.
8. A. Chapman, "Hacking into internet connected light bulbs," *ConCon Blog*, July 4, 2014.
9. S. Checkoway and others, "Comprehensive experimental analysis of automotive attack surfaces," in *Proceedings of the 20th USENIX Security Symposium*, 2011.
10. C. Cimpanu, "Police body cameras shipped with pre-installed Conficker virus," *Softpedia*, November 15, 2015.
11. C. Cimpanu, "RSI videofied security alarm protocol flawed, attackers can intercept alarms," *Softpedia*, November 30, 2015.
12. C. Cimpanu, "Company takes two years to remove hard-coded root passwords from IoT thermostat," *Softpedia*, February 8, 2016.

13. C. Cimpanu, "Vulnerability in 70 CCTV DVRs traced back to Chinese firm who ignores researcher," *Softpedia*, March 23, 2016.

14. A. Connolly, "Thousands of Android and iOS apps are still vulnerable to the FREAK bug," *The Next Web*, March 18, 2015.

15. L. Constantin, "Researchers show that IoT devices are not designed with security in mind," *PC World*, April 7, 2015.

16. L. Constantin, "Cisco patches serious flaws in cable modems and home gateways," *CSO Online*, March 10, 2016.

17. L. Constantin, "FortiGuard SSH backdoor found in more Fortinet security appliances," *CSO Online*, January 22, 2016.

18. A. Costin and others, "A large-scale analysis of the security of embedded firmwares," in *Proceedings of the 23rd USENIX Security Symposium*, 2014.

19. E-ISAC, "Analysis of the cyber attack on the Ukrainian power grid," SANS Industrial Control Systems, March 18, 2016.

20. D. Fisher, "Default SSH key found in many Cisco security appliances," *ThreatPost*, June 25, 2015.

21. D. Fisher, "Plaintext credentials threaten WRLE wind turbine HMI," *ThreatPost*, June 17, 2015.

22. J. Foster, "Someone gained access to private PLQ meetings, very easily," *CJAD News*, June 17, 2016.

23. T. Fox-Brewster, "Hacker says attacks on 'insecure' progressive insurance dongle in 2 million US cars could spawn road carnage," *Forbes*, January 15, 2015.

24. D. Goodin, "Backdoor in mission-critical hardware threatens power, traffic-control systems," *Ars Technica*, April 25, 2012.

25. D. Goodin, "Crypto weakness in smart LED lightbulbs exposes Wi-Fi passwords," *Ars Technica*, July 7, 2014.

26. D. Goodin, "Asus lawsuit puts entire industry on notice over shoddy router security," *Ars Technica*, February 23, 2016.

27. D. Goodin, "Samsung Smart Home flaws let hackers make keys to front door," *Ars Technica*, May 2, 2016.

28. T. Goodspeed and others, "Packets in packets: Orson Welles' in-band signaling attacks for modern radios," in *Proceedings of the 5th USENIX Conference on Offensive Technologies*, 2011.

29. A. Greenberg, "Hackers remotely kill a Jeep on the highway—with me in it," *Wired*, July 21, 2015.

30. N. Heninger and others, "Mining your Ps and Qs: Detection of widespread weak keys in network devices (extended version)," in *Proceedings of the 21st USENIX Security Symposium*, 2012.

31. S. Higginbotham, "When it comes to smart home security, cameras are the worst," *Gigaom*, February 11, 2015.

32. K. Hill, "This guy's light bulb performed a DoS attack on his entire smart house," *Fusion*, March 3, 2015.

33. I. R. Jenkins and others, "Short paper: Speaking the local dialect: Exploiting differences between IEEE 802.15.4 receivers with commodity radios for fingerprinting, targeted attacks, and WIDS evasion," in *Proceedings of the 2014 ACM Conference on Security and Privacy in Wireless and Mobile Networks*, 2014.

34. D. Kaminsky and others, "PKI layer cake: New collision attacks against the global X.509 infrastructure," in *Financial Cryptography*, 2010.

35. J. Kirk, "Netcore, Netis routers at serious risk from hardcoded passwords," *InfoWorld*, August 26, 2014.

36. J. Kirk, "Researchers find same RSA encryption key used 28,000 times," *ITworld*, March 16, 2015.

37. J. Kirk, "Tens of thousands of home routers at risk with duplicate SSH keys," *Network World*, February 18, 2015.

38. J. Leyden, "Samsung smart fridge leaves Gmail logins open to attack," *The Register*, August 25, 2014.

39. D. Maas, "The world's most dangerous search engine," *San Diego CityBeat*, February 6, 2013.

40. C. Masone and S. W. Smith, "ABUSE: PKI for real-world email trust," in *EuroPKI '09 Proceedings of the 6th European Conference on Public Key Infrastructures, Services and Applications*, 2009.

41. R. Merrriam, "Software update destroys $286 million Japanese satellite," *Hackaday*, May 2, 2016.

42. C. Neagle, "Smart home hacking is easier than you think," *InfoWorld*, April 3, 2015.

43. J. Newman, "Internet-connected Hello Barbie doll can be hacked," *PC World*, December 7, 2015.

44. J. C. Norte, "Hacking industrial vehicles from the internet," *Jose Carlos Norte Personal Blog*, March 6, 2016.

45. M. Reel and J. Robertson, "It's way too easy to hack the hospital," *Bloomberg Businessweek*, November 2015.

46. L. Sassaman and others, "Security applications of formal language theory," *IEEE Systems Journal*, 2013.

47. R. Shapiro and others, "Identifying vulnerabilities in SCADA systems via fuzz-testing," in *Critical Infrastructure Protection V*, Volume 367, 2011.

48. M. Smith, "Peeping into 73,000 unsecured security cameras thanks to default passwords," *Network World*, November 6, 2014.

49. S. Smith and J. Marchesini, *The Craft of System Security*. Addison-Wesley, 2008.

50. S. W. Smith, "Pretending that systems are secure," *IEEE Security and Privacy*, November/December 2005.

51. S. W. Smith, "Room at the bottom: Authenticated encryption on slow legacy networks," *IEEE Security and Privacy*, July/August 2011.

52. S. W. Smith and R. Koppel, "Healthcare information technology's relativity problems: A typology of how patients' physical reality, clinicians' mental models, and healthcare information technology differ," *Journal of the American Medical Informatics Association*, June 2013.

53. S.W. Smith and others, *Mismorphism: a Semiotic Model of Computer Security Circumvention (Extended Version)*. Dartmouth Computer Science Technical Report, March 2015.

54. P. Svenda and others, "The million-key question—Investigating the origins of RSA public keys," in *Proceedings of the 25th USENIX Security Symposium*, 2016.

55. todb, "Advantech EKI Dropbear authentication bypass," *Rapid7 Community*, January 12, 2016.

56. J. Ullrich, "Did you remove that debug code? Netatmo Weather Station sending WPA passphrase in the clear," *SANS ISC InfoSec Forums*, February 12, 2015.

57. C. Vallance, "Car hack uses digital-radio broadcasts to seize control," *BBC News*, July 22, 2015.

58. Z. Whittaker, "Widely-used patient care app found to include hidden 'backdoor' access," *ZDNet*, May 27, 2016.

59. V. Zhang, "High-profile mobile apps at risk due to three-year-old vulnerability," *TrendLabs Security Intelligence Blog*, December 8, 2015.

5

Names and Identity in the IoT

The scale of the IoT vision outstrips current authentication technology—so we will need to think of something new.

In the standard vision of the IoT, massive numbers of things are talking to other things. For this talking to be meaningful, the listeners need to know who the talkers are. Was it really my car's ECU that just told my car's brakes to engage? Was it really my washing machine that just asked Google for details of my calendar, or that just told my utility company that the machine is using a less power-hungry washing algorithm?

Thanks to the permeable nature of networked communication, impersonation is always a concern. However, thanks to the IoT's wide physical distribution and intimate connection to reality, impersonation in the IoT may have serious consequences—maybe it was the digital radio receiver, fooled by remote hackers, that was pretending to be my car's ECU. Making things even more complicated is that a "globally unique name," even if such a thing existed, may not suffice for the listener; when someone claiming to be a police officer knocks on your door, you don't care about the officer's Social Security number and DNA fingerprint.

This chapter explores the challenges of naming and attributes in IoT-scale populations—and the corresponding challenges of effective techniques (cryptographic and otherwise) to provide this data.

Who Is That, Really?

The basic challenges of identity and authentication showed up in Chapter 4, which discussed some patterns in which such things have gone wrong already.

Usually, these concepts appear in the context of electronic communication. When Alice's machine receives a message *M*, it's useful for Alice to know its provenance. Usually this provenance consists of two elements:

Identity
: Who allegedly sent this message?

Authentication
: Did that party actually send it?

Sometimes, the concepts extend to the communication channel: how Alice's machine can know that it's really Bob's machine on the other side of the pipe, so that subsequent messages she sends down the pipe are caught by Bob's machine and not something else.

In the IoC, a common example of this is the dance done between your browser and a remote server when you want to buy something from Amazon: your browser uses PKI (described in more detail later in this chapter) to set up a cryptographic channel whose other end is (one hopes) the Amazon server, with no eavesdroppers along the way. Another example, in the IoT and the IoC, would be software updates: one hopes that a device verifies that some alleged software update actually came from the appropriate party, rather than an impostor, before accepting and installing it. (Unfortunately, as Chapter 4 discussed and as the Ukrainian power grid observed in December 2015, that isn't always the case.)

More straight-ahead IoT examples include whenever a smart thing accepts a command to act on the physical world ("open that door," "turn that generator off," "change the heart rate on that pacemaker") or report on the state of the physical world ("there's plenty of room in that storage tank, so keep pumping in the fuel"). As these examples suggest, both kinds of messages—commands and responses—might have serious consequences if forged by a malicious party intent on causing damage.

Of course, as discussed in many places (including Chapter 9 in my security textbook [15]), there's more to the picture than simply that the message was really sent by Bob. For example, usually the context in which Bob sent the message matters; otherwise, an adversary could fool Alice simply by replaying an old message from Bob. Designing the dance correctly is tricky.

BEYOND BITS

The preceding discussion framed authentication in the context of electronic communication, because that's the framework in which it usually arises in computing (and also because it's easier to use standard cryptographic tools then). But other contexts also apply.

One such context already arises in the IoC when one of the parties is a human, with no computational enhancement. How does Alice's computer authenticate that Alice is at the keyboard? Here's where a standard security textbook would start talking about different factors for authenticating humans ("something you know," "something you have," "something you are") and the increased rigor of using multifactor authentication. However, unfortunately, there's another direction to human–computer authentication that receives less attention. How does Alice authenticate that it's really her computer, or really the genuine "enter your password now" window from her browser, and not a clever spoof from a malicious site?

In the IoT, a context that will likely become more critical is authentication of *physical proximity* rather than strings of bits. Which electric car just drove by the charging station? Which health monitor bracelet should report to Alice's phone? Which smart meter should Alice's washing machine talk to?

A body of work stretching from the seminal "Resurrection Duckling" paper [18] and continuing through the recent "Wanda" work by Tim Pierson [19] explores allowing end users to use physical proximity to assert an ontological association.

AUTHORIZATION

The persnickety reader might point out that in a closer reading of this problem, what matters is not the "identity" (e.g., name) of the party, but whether they are the "right" party. For a human-space example, when hospital patient Alice is about to receive an injection from someone dressed in white, what matters to Alice is not the name of this person, but that he or she is a genuine clinician. As Chapter 4 hinted, this problem can be messy even in the IoC.

The Standard Cryptographic Toolkit

Chapter 4 briefly introduced the basic concepts of cryptography. This section will go into a bit more detail, since cryptography provides the bulk of the toolkit used to try to solve these problems in practice (in computing settings). For more detail, however, the reader should consult a standard security textbook, such as Chapters 7 and 10 in mine [15].

THE SOMEWHAT IMPOSSIBLE

Casual discussions of applied cryptography often refer to impossibilities. "If Alice does not have the key K, then she cannot calculate $y = D_K(X)$." One may wonder: what exactly does "cannot" mean?

The answer is both satisfyingly precise and annoyingly ambiguous. Building on early 20th-century work in logic, computer science has developed nice mathematical formalisms for the concept of possible computation:

Computability
: Can a computational algorithm actually exist for a given problem?

Size and Complexity
: Thinking of a computational problem as a family of instances, one for each possible value of its input parameters, we can define its size in terms of the size of the input parameters. For a particular algorithm solving a computational problem, we can precisely talk about how the sizes of the necessary computational resources (e.g., time or space) grow with the size of the problem instance.

Complexity theory nicely factors out the differences due to low-level details of particular computing platforms. It lets us talk about *upper bounds* (the maximum resources necessary to solve some problem) and, less often, *lower bounds* (the minimum resources required). It also lets us talk about *reduction*: sometimes, whenever we have an instance of a problem P_1, we can turn it into an instance of P_2. Reduction thus lets us connect the complexity of one problem to that of the other; for example, P_1 can be no harder than P_2 plus the transformation. (And yes, a full treatment would require thinking about more general variations, such as solving P_1 using subroutine calls to a P_2-solver.)

Complexity lets computer scientists talk about *intractability*. A problem is intractable when, even though it can be computed, the minimum resources necessary grow far too quickly as problem instance size grows—so once the instance becomes nontrivial, it would take more time than remains in the known universe's lifetime to finish the calculation.

When computer scientists talk about "impossibility," they're usually using this framework. However, the framework gives a range of definitions:

Impossible

- The problem is in fact not computable.

Practically impossible

- The problem is computable, but provably intractable.

Believed practically impossible

- The problem is computable and not yet proven intractable, but has been shown to be at least as hard as a large family of problems everyone believes are probably intractable (e.g., *NP-hard*).

Hopefully practically impossible, with some evidence

- The problem is computable and, while not known to be NP-hard, is at least as hard as another problem (e.g., factoring) whose difficulty is unknown but which everyone suspects is hard.

Hopefully practically impossible, with no evidence

- The problem is computable and not known to be as hard as a suspected-hard problem, but it looks difficult.

The reason the "annoying ambiguity" comes in is that, in practical cryptography, the "impossibility" definition is usually one of these last two: "hopefully practically impossible."

SYMMETRIC CRYPTOGRAPHY

A basic way to approach it is to think of it as a pair of transformations: E, taking plain text to *ciphertext*, and D, doing the reverse—so $D(E(x)) = x$ for any x. (Think "encipher" and "decipher," as the perhaps more natural terms "encrypt" and "decrypt" have more specialized meanings to some subcommunities.) However, for this to be useful, the system needs the additional constraint that at least one of these functions (if not both) requires a special privilege to be able to carry it out. This privilege gets embodied as an additional numeric parameter called a *key*.

For most of human history, there was only *symmetric* cryptography, where the key necessary to perform E was the *same* key necessary to perform D. If $y = E(k, x)$, then only a party who knew k could take y back to x via $D(k, y)$.

Common symmetric algorithms include *DES* (the US standard for a few decades, now retired) and *AES* (the new US standard).

The impossibility criterion here (that one must know k to take $E(k, x)$ back to x) is generally the weakest: "hopefully impossible." For all practical cryptographic schemes, an adversary can certainly perform a forbidden transformation by *brute force*: trying all possible keys. Hence, discussions of crypto schemes usually talk about *key length*: the longer the key, the longer a successful brute-force attack would take. Deployments ended up hoping that the keys were long enough that this type of attack was essentially infeasible.

However, history provides some cautionary tales. Consider last century's DES: a widely used security textbook still said it was safe to use even after a team coordinated by the Electronic Frontier Foundation (EFF) had built and demonstrated a machine that could brute-force crack its 56-bit keys. Another noteworthy trend was the emergence (in the public world) of successful *side-channel* attacks. The computers that perform the cryptographic computation are themselves machines in the physical world, and it turned out that by measuring properties of these machines (such as the time they took or the power they consumed), one could often learn the key they were using. (I personally saw a commercial DES device reveal its key with only *one* computation: when we looked at the power consumption, there were 56 spikes, some short, some long.)

Perhaps the most natural use of symmetric cryptography is to protect messages for secrecy. If Alice wants to send a message M to Bob and have confidence that no one else can read it during transit, she can instead send $E(k, M)$—as long as she and Bob already know k. Symmetric cryptography can also be used for authentication. For a simple and occasionally dangerous example:

1. Bob might send Alice some random x.
2. Alice encrypts it with her key k and returns it.
3. Bob decrypts the returned value and checks that he gets x back.

Again, Alice and Bob need to share k first.

Another variation is *message authentication*. Again, for a simple and occasionally dangerous example:

- Alice and Bob share a secret key k.
- Alice wants to send a message M to Bob, but have Bob have assurance that Alice really sent it.
- So, she uses her secret key and favorite symmetric cryptosystem to calculate some short message authentication code based on M. For example, maybe she calculates the full $E(k, M)$, but then uses the last 16 bytes as the MAC.
- Alice sends the message and the MAC to Bob.
- Bob then calculates the MAC on the message he received, and compares it to the MAC that he received. If they match, then he has high confidence that the sender of the message also knows the secret key k.

PUBLIC KEY CRYPTOGRAPHY

As the previous examples kept repeating, use of symmetric cryptography requires that the parties involved share a key. This requirement can be problematic. One reason is logistics: if Alice and Bob need cryptography to set up a secure communication channel, then what channel do they use to set up the shared key? Another is scalability: if Alice wants to talk to n parties, she will need n different secret keys; if n players need pairwise communication, the system needs about n^2 different keys. Yet another reason is semantic: a shared key is useful only among the parties who share it. When Bob receives a message M from Alice, he can conclude it came from Alice by verifying its MAC with the key he shares with her. But he can't use this MAC to prove to anyone else that Alice said M.

Public key cryptography (also known as *asymmetric* cryptography) provides ways to overcome these limitations.

From the dawn of time to the age of disco (as I like to point out each time I teach about this material), humanity only had symmetric cryptography. In the 1970s, the idea emerged of replacing the single key with a *keypair* k_e, k_d, such that enciphering under k_e is inverted by deciphering under k_d. At least one of these keys cannot be derived from knowing the other—hence the term "public key," since this latter one can be made public without harm. For standard public key cryptosystems such as RSA, the impossibility criteria is "probably impossible": deriving the nonpublic key from the public information appears to be as hard as factoring.

Public key cryptography can do the same things symmetric cryptography does, except now Alice and Bob no longer need to share a secret key:

Encryption
: If Alice wants to send message M to Bob and keep the contents secret from anyone else, she need only encipher it with Bob's public key. Only someone knowing Bob's private key can then transform this back to M.

Message authentication
: If Alice wants Bob to know that she sent message M, she can *digitally sign* it by doing a transformation with her private key. Bob—or anyone else—can verify that this signature matches M using Alice's public key.

Entity authentication
: Alice can authenticate herself to Bob by digitally signing a random challenge he provides.

How do parties know the public keys of other parties? Read on.

As before, these examples are just simple sketches. Real-world uses of public key cryptography usually have more complexity; encryption will also use symmetric cryptography (discussed previously), signatures will also use cryptographic hash functions (described shortly), and both will likely use various padding functions.

PUBLIC KEY INFRASTRUCTURE

Initially, using public key cryptography brings the number of keys a device needs to know down to n: its own keypair, and the public key of each other device it wants to communicate with. However, the ability of public key cryptography to enable digital signatures—a party can use its private key to sign an assertion verifiable by anyone knowing its public key—brings the number down to two: a device only needs to know its own keypair and the public key of the *trust root* it trusts to sign assertions saying what the public keys are of the other devices with which it needs to work. In either case, the number of *secrets* a device needs to know is exactly one: its own private key. This constraint limits the damage that compromise of a device can cause.

Public key infrastructure is the catch-all term for the mechanics necessary to establish, maintain, and distribute these assertions (*certificates*) and keypairs. Typically, PKI includes components to solve these problems:

- How does a keyholder obtain a certificate from a trust root?
- How does a relying party decide who its trust roots are?
- How does a relying party get its hands on a *path* from a trust root to a particular certificate?
- What exactly should a relying party conclude from discovering a path with each link apparently signed properly?
- What do we do when the assertion a certificate makes (e.g., "X has public key E_x") is no longer true and needs to be *revoked*?

X.509 is the family of standards and techniques (e.g., [9]) that has come to dominate the way most PKI is done in practice, although other rivals (e.g., [5, 14]) surface now and then, as do regular critiques (e.g., [8]). X.509 allows for *identity certificates* (assertions about the name of a keyholder) as well as *attribute certificates* (assertions about some more general properties), although use of the latter is more experimental. X.509 (and other PKI approaches) also defines some semantics for what should be concluded from a chain of certificates: if Alice issues a signed assertion about Bob, and Bob issues one about Cathy, what conclusion should Dave draw if he sees both assertions?

CRYPTOGRAPHIC HASHING

Another item in the standard toolkit is *cryptographic hash* functions—essentially, encryption that cannot be decrypted. (In the context of security, these are often just called *hash* functions; the "cryptographic" is implied.)

A hash function H takes an arbitrarily long string x down to some fixed-length strength $H(x)$, but with the property that this transformation cannot be inverted: given a z, one cannot find a y such that $H(y) = z$. (The impossibility criterion is usually "hopefully impossible.") This impossibility still holds even if one is given a number of other hints, such as a y' that hashes to something close to z, or the first half of a y that hashes to z.

An early application of such *one-way* functions was for checking whether an entered password was correct (do the one-way transformation, then compare that to some stored value) without allowing an adversary who gets hold of the file of transformed passwords to recover the originals.

SHA-2 is an example of a hash function currently in use.

One use of hashing is to shorten digital signatures: instead of applying her RSA private key to a whole message M, Alice will first calculate $H(M)$ and then apply her private key to that. Another is to almost do an asymmetric proof-of-knowledge: if Alice knows a secret x, she can publish $z = H(x)$ without compromising its secrecy—and then later prove to Bob she's Alice by exhibiting x (Bob can hash it and check). Of course, this only works once, but schemes exist where Alice can extend the lifetime by publishing a cascaded value like $H(H(H(H(x))))$ and unraveling it one step at a time.

Hashing can also be used to tie values together. If block B_2 contains a hash of block B_1, then changing something in B_1 invalidates the hash in B_2. This dependency can extend to a chain: if we have a sequence $B_1, B_2, \ldots, B_N$ with each B_{k+1} containing $H(B_k)$, then someone examining B_N can tell whether any previous B_i has been altered (see Figure 5-1).

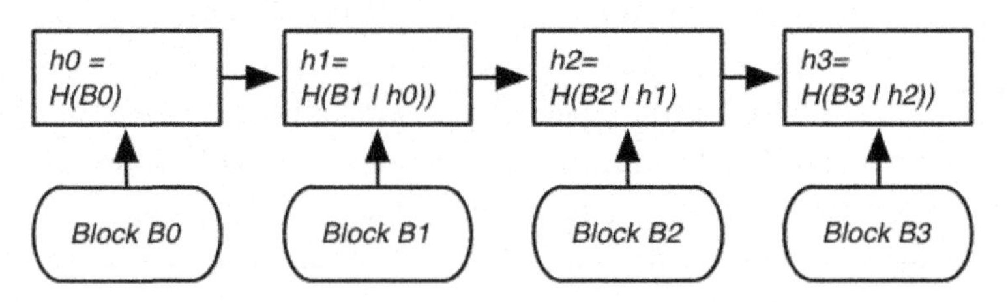

Figure 5-1. If we have a chain of data blocks, having the hash of each block depend on the previous hash means someone examining B_N can tell whether any previous B_i has been altered.

We can also tie this dependence in with the "hard to invert, even with hint" property described earlier. For example, if we have four (digital) documents $D_0, \ldots, D_3$, we can organize them into a binary *Merkle* tree, walk the tree from the bottom to the top, and label each node with a hash of its children (as Figure 5-2 (A) shows). If we then publish the root in the *New York Times*, we can prove that any of these documents existed before that publication by (as Figure 5-2 (B) shows) exhibiting the right sequence of hints that show how that document contributed to the published root. In fact, Stuart Haber and colleagues created a company that did just this—and for good measure, had each published root depend on the previously published root, to make it that much harder to forge history.

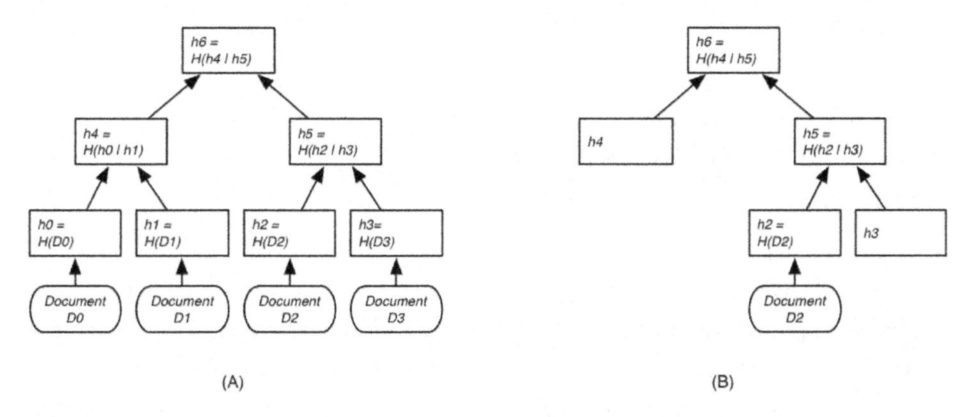

Figure 5-2. (A) A Merkle tree puts the hashes of 2^k documents as leaves in a binary tree and labels each non-leaf with a hash of the concatenation of its children. (B) The owner of a document can prove it contributed to the root by exhibiting the correct sibling hashes.

In its general presentation, hashing is a public operation: anyone can perform *H*. A variation is a *keyed hash*, where the hash operation requires a secret key. The accepted standard here is *HMAC*: a way of building a keyed message authentication code function from a public hash function.

THE PRICE TAG

Public key cryptography sounds much more attractive than symmetric key, because it eliminates the hassles and restrictions of all these shared secrets. But these benefits come with a price: it's significantly more computationally intensive than either symmetric key cryptography or hashing. For tiny devices and for devices communicating over constrained channels, these costs can be problematic—and the IoT will have plenty of both. (The use of newer public key schemes such as *elliptic curve cryptography* reduces but does not eliminate this problem.)

It's worth pointing out that, as a consequence of these engineering constraints, many computing infrastructures use symmetric key cryptography. Historically, automatic teller machines (ATMs) used a scheme based on symmetric keys to validate customer transaction requests, although that is being supplanted by public key schemes. Cellphone authentication essentially works by having each device share a symmetric key with a master server associated with the service provider, and then having provider-to-provider bridging arrangements. However, again, as telephony merges with the internet and the World Wide Web, we're seeing more adaption of public key techniques.

The Newer Toolkit

The tools just described are standard items one sees in the cryptographic toolkit for the IoC. However, a few less-standard items (some new, some less new) are being proposed for the IoT specifically.

MACAROONS

One such tool is the *macaroon*, from Google (e.g., [3]).

Suppose we have a sequence of IoT computing machines C_0, C_1, C_2 where order corresponds with size and authority. For example, C_0 might be a big server at a local electric power utility, C_1 a smart meter that utility installed at a consumer's house, and C_2 a smart appliance that consumer buys—an appliance that offers some useful features with regard to power consumption. In a natural authority flow, the utility blesses the smart meter and the smart meter blesses the appliance. In this scenario, it might be useful for the utility server C_0 and appliance C_2 to be able to establish mutually authenticated communications: for instance, so the appliance can report usage information to the server, and the server can ask the appliance to use less power when the grid is under stress. However, this can be tricky if the utility and the appliance have never met.

PKI provides a natural solution to this problem. Each party has a keypair. C_0 signs an assertion about C_1, and then C_1 signs one about C_2. The pair of assertions could then suffice to tell C_0 that the holder of the C_2 keypair is in fact that appliance at that house; standard PK techniques would then allow C_0 and C_2 to set up a secure, mutually authenticated communication channel.

However, as noted earlier, public key cryptography is expensive. The idea of macaroons is to use the more lightweight HMAC to achieve similar functionality.

Figure 5-3 (A) shows the basic construction of a macaroon. Computer C_0 has a secret k_0. It generates a random nonce n_1 and appends some information (*caveats*); it uses k_0 to then generate the HMAC k_1 of these items. We now have the public macaroon (consisting of the nonce and the caveats) and the private k_2. If C_0 then passes the macaroon and k_1 (over a secure channel) to machine C_1, then C_1 shares a secret with C_0. So later, C_1 can interact over an open channel with C_0, announce its macaroon, and then they both can use the shared secret k_2 to set up a mutually authenticated channel.

So far, this doesn't buy us much; C_0 could do the same thing by storing a huge database of pairs of macaroon text and corresponding key. But we can start to see the power when we start chaining macaroons together. For example, consider the scenario in Figure 5-3 (C). After C_1 receives its macaroon and key k_1, it can use these items to generate a second macaroon and key k_2, and implant these in computer C_2. Now C_2 can mutually authenticate with C_0—even though C_0 may never have heard of C_2 before—simply by presenting its macaroon, which can be thought of as a publicly readable ticket that enables C_0 to derive the shared secret k_2.

As Figure 5-3 (B) shows, the public macaroon corresponds roughly to a PKI certificate, saying that according to the creator of the macaroon, the party knowing the matching secret key has the properties specified in the caveats. As Figure 5-3 (D) shows, macaroon chaining is similar to certificate chaining. The downsides are that macaroons cannot be verified by the general public, but only by parties higher up in the macaroon chain (who know or can derive the generating secret), and that they don't have the asymmetry and nonrepudiation of keypairs: anyone higher in the chain can also do things with a party's secret key. However, the upside is that instead of having expensive operations such as RSA's modular exponentiation on massive integers, we have lightweight hashing and symmetric cryptography.

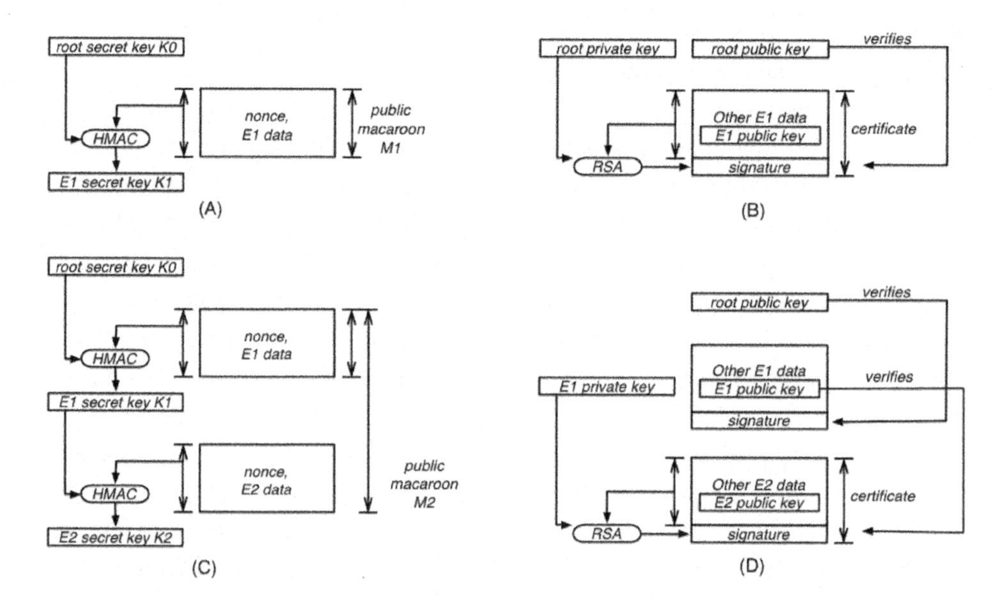

Figure 5-3. (A) With macaroons, an entity exhibits a macaroon to prove its identity and uses this shared secret to authenticate itself to a relying party knowing the trust root's secret key. (B) With PKI, an entity exhibits a certificate to prove its identity and uses its private key to authenticate itself to any relying party knowing the trust root's public key. (C,D) Both approaches can be chained.

BLOCKCHAINS

Blockchains are another tool currently receiving much attention for many applications, including the IoT.

Blockchains burst into public awareness with the emergence of *Bitcoin*, a set of protocols and related infrastructure for *cryptocurrency*. The idea of cryptocurrency is to use the magic of cryptography and such to mint "coins" that are just strings of bits, but which have the properties of cash: that is, a coin can only be spent once. Cryptocurrency schemes usually have a strong antiauthoritarian bent, and so value decentralization. The Bitcoin protocol, published in 2008 by the probably pseudonymous Satoshi Nakamato [12], addresses this double-spending problem by using hash chaining (and some other tricks) across a decentralized peer-to-peer network to build a hard-to-forge immutable record of transactions—so if someone tries to cheat by spending money he or she has already spent, the entire network can testify.

The basic idea is to represent each transaction as a data *block*. The blocks are chained together in a master sequence. Each block in this chain contains a hash of the previous block (Nakamato cites Haber's timestamping work) so that modifying history would require rewriting all the blocks following the modification. The chain is shared universally among nodes in a peer-to-peer network, who use techniques such as "go with the longest chain" to establish consensus in case of disagreement. To keep a malicious party from appending a long stream of blocks to the chain in order to suppress a legitimate transaction an honest party is trying to add (since the malicious extension would then be the "longest chain"), the system uses a *proof-of-work* mechanism to limit the rate at which blocks can be created. Each block needs to contain a special value *x* whose hash value begins with a specific number of zero bits (and this number differs and increases for each block in the chain). By the impossibility assumptions underlying hashing, the only way for the would-be constructor of a block to find such an *x* is to expend computing power doing brute-force searching.

The term "blockchain" has come to denote this general idea of using a distributed network to maintain a hashed chain of blocks whose creation is rate-limited. Its value for cryptocurrencies (and other applications) is that it provides a decentralized, immutable *ledger* of a transaction stream:

- All nodes agree on the current history of valid transactions.
- All nodes can then witness that a new transaction is valid before appending it to the history.

We can look at currency transactions as transitions in a state machine:

- At any given point, the state of the system is who has how much currency.
- A transaction is valid for a given state when the state satisfies its precondition: the spender actually has the money he or she is spending in the transaction.
- Accepting the transaction as valid thus changes the state: money has shifted.

This view—transactions with preconditions modifying state—can quickly generalize to any kind of computation, and blockchain projects have done this. Transactions can be written as *smart contracts* or *chaincode*, firing automatically when their preconditions are satisfied. The *Ethereum* blockchain project even adds a Turing-complete programming language to its transaction model. Other blockchain extensions include adding permissions/access control to blocks and their fields, although (at first glance) this would seem to reduce effectiveness by reducing the population of witnesses and verifiers. Many players—including large corporations and the Linux Foundation—are now exploring blockchains for applications other than cryptocurrency.

As the start of this section mentioned, one such application is the IoT. The motivation is that blockchains in the IoT may hit several "sweet spots":

- Projections of exponentially growing numbers of things suggest that the backend server and communication channels between the things and the backend may become bottlenecks. Replacing the centralized backend with a decentralized ledger eliminates this problem.
- The IoT things themselves are already a distributed population, interacting.
- The IoT may involve so many players—consumers, vendors, service providers, utilities, retail stores, etc.—that relying on any kind of centralized authority may be impractical.

At CES 2015, IBM and Samsung even demonstrated using the Ethereum blockchain and smart contracts to enable an IoT-aware washing machine to order detergent when necessary, to negotiate sharing power usage with the household TV, and to arrange warranty service. IBM touts this approach to the IoT of autonomous peers as "device democracy."

As happens with any hot, potentially disruptive technology, blockchains (in the IoT or elsewhere) are currently going through a bit of churn. Some enthusiasts insist that permissioned blockchains don't really count as blockchains. In the cryptocurrency world, the Mt. Gox bitcoin exchange collapsed in 2014 (with close to half a billion dollars missing, perhaps gone to hackers); in June 2016 the Ethereum-based DAO lost about $60 million, apparently to hackers, and in August the Bitfinex exchange suffered a high-profile attack. Stay tuned.

PUFS

Having devices use cryptography with secret or private keys requires that devices keep their keys secret—since if an adversary learns a device's special key, then the adversary can start impersonating the device. Historically, protecting secrets inside computational devices has been a cat-and-mouse game, with each side—attacker and defender—successively coming up with innovations; see my survey article [17] for more information.

Simply by vastly increasing the number of computation devices, the IoT risks making this problem hard. That the computational part of IoT devices may need to be inexpensive will make it harder still. As a consequence, some researchers are considering using *physically unclonable functions* (PUFs) to help address this problem in the IoT.

PUFs emerged from Srini Devadas's lab at MIT in 2002 [6]. The basic idea is to "magically" embed a nonpredictable number in a computational device, such that the device can make computational use of this number, but an adversary who tries to access the number in some other way (e.g., by prying off the lid and probing) will not be able to read it. The number is accessible only when it is being used in the circuit itself—it's a physical artifact of the circuit in operation.

This number and the computations a device does with it can thus be used as a foundation for cryptographic identification and other operations.

Of course, PUFs are not really made by magic. Rather, the technology depends on subtle physical and electrical properties and manufacturing variations—which is how they aspire to resist the standard FIPS 140 attack of "opening up and looking, and building a clone." The adversary might look but not see; the adversary might know but cannot clone. At least, that's the design. As someone who worked on the first full top-level FIPS 140–validated device, I can also testify to another potential advantage of PUFs for the IoT: straight-on tamper resistance/detection/response is expensive and tricky, and perhaps not economical for devices that need to be cheap and live in environments more extreme than machine rooms in data centers. PUFs may overcome these limitations.

They have potential, but whether PUFs will stand the test of time as a robust but inexpensive way to hide secrets remains to be seen.

ADDRESSES AND NAMES

"Connection to Other Computers" on page 42 discussed how networking protocols such as IPv6 can give us the ability to give each IoT thing a unique address and (if everyone behaves) send data there. However, this doesn't itself solve the identity problem.

What does the name *mean*? An IP address itself doesn't carry much semantic information; that's why people use hostnames such as "www.cs.dartmouth.edu" rather than IPv4 addresses like "129.170.213.101"—and even then, the semantics in the hostname (e.g., the web server of the "CS" part of "dartmouth," which is some kind of educational institution) follow from convention, and are often misinterpreted by humans. This mapping is determined by the Domain Name System (DNS), and securing the DNS is a continuing challenge. DNS Security Extensions (DNSSEC) is one ongoing attempt to build a giant PKI to make sure the mapping is authentic.

How do we actually route data to a particular thing? The legacy IoT uses the Border Gateway Protocol (BGP) to figure out how to get data to the right IP address—but BGP is widely lamented for being built on the assumption that the big nodes in the internet actually tell the truth when they share routing information with one another. As a consequence, malevolent actors can trick parts of the internet into sending data to the wrong places. Addressing the problem of routing security is another area of ongoing research.

IoT Challenges

The previous section discussed toolkits that might help with identity and authentication in the IoT. What are the challenges?

ONTOLOGIES OF ASSOCIATION

In the IoT, what does a thing need to know about another thing?

- What attributes do listeners need to know?
- Who is in a position to witness to these attributes? (When does it become *my* washing machine?)
- When will the binding change? (What if I sell the car or move out of the apartment?)

Traditional cryptographic approaches to identity and authentication focus on binding a secret key or public/private keypair to the keyholder's identity, which is implicitly assumed to be a well-defined, relatively static thing—such as, in PKI, an individual's full name or email address, or the hostname of a public web server.

However, in the envisioned IoT, the relevant properties of the keyholder are not just the device's identity ("this is a meter made by ACME"; "this is a refrigerator made by GE") but its *context* ("this is a refrigerator in the apartment rented by Alice, who buys power from X"). This context information will not necessarily be known until device installation, and it may change dynamically. (What if Alice sells her fridge on Craigslist or sublets her apartment to Bob? What if repair person Cathy replaces Alice's meter?). This information may also not be particularly simple. (What if Alice's landlord owns many apartment buildings, and changes power vendors to get a better rate?)

If a cryptographic infrastructure is going to enable relying parties to make the right judgments about these smart devices, this additional information needs to be somehow available. We can try to modify a traditional identity-based PKI to attest to these more dynamic kinds of identities, or we could try instead to adapt the largely experimental world of attribute certificates (e.g., [4]) to supplement the identity certificates in the IoT PKI. Or we might go with the more lightweight but more constrained macaroons. Any of these approaches breaks new ground. Alternatively, we can leave the identity PKI in place and use some other method of maintaining and distributing this additional data; this requires supplementing our scalable PKI with a nonscalable database.

In any of these approaches, we also need to think about who is authorized to make these dynamic updates. Who witnesses that Alice has sold her refrigerator? Thinking about this organizational structure of smart grid devices also complicates the revocation problem. If we can't quite figure out who it is that speaks for where a device currently lives, how will we figure out who it is who is authorized to say it has been compromised?

ONTOLOGIES OF INTERACTION

In the previous section we considered the granting of names. Another angle is to consider how names must be used. Which thing needs to talk to which other thing, and how often? How broad or narrow are the patterns of communication that need to carry this naming and attribute data? Every traffic light in the US does not need to talk with my washing machine—but they *might* need to talk with my car.

The standard argument that public key technology makes it easier to have large populations talk to one another starts losing impact if we don't need to solve that fully general problem. For example, in the power grid, name management (and the corresponding management of keys for identity and authentication) is much simpler for legacy high-voltage transmission lines—since they don't change endpoints every often—than for future electric cars which may charge as they drive. If we have an IoT structure where smart things in a household talk to one another but must go through a proxy before leaving the house, then it might be feasible just to have the proxies visible.

Indeed, this problem space has many dimensions:

- What needs to be known about an IoT thing
- Who is in a position to testify to this name or attribute
- Who needs to know this information

PKI AND LARGE POPULATIONS

As this chapter has noted, public key technology may seem like a natural solution to this large-scale IoT problem. However, it comes with engineering challenges. We'll start with some of the issues relevant to PKI for any large-scale population —and the IoT, with billions of entities, is likely to be larger than any computing population we have tried so far.

Trust roots

In the textbook view of PKI, a single trusted entity acts as the universal *certification authority*. This one party issues all certificates; this one party's public key is the only trust root any relying party ever needs to know. Unfortunately, even in the PKIs that have emerged so far (relatively small, compared to the IoT), this simplifying property has failed to hold. For reasons logistical, economic, and otherwise, multiple CAs emerge serving various parts of the population. Consider: as of this writing, the population of SSL-protected web servers are served by over 100 different trust roots—and the IoT will have far more devices in far more homes and businesses than the IoC currently has SSL-protected web servers. So, for the IoT grid, we will have to either figure out how to solve a problem no one's solved yet—having one entity sign certificates for a vast population —or deal with the reality of myriad trust roots.

Having myriad trust roots raises the question of how these roots should be organized. One might imagine a strict hierarchical tree, where higher-level roots certify lower-level ones, and the lowest-level roots certify devices. One might also imagine a system where peer roots have their own user populations, but *cross-certify* one another. ("My users can trust R_2 to talk about devices in subpopulation P_2.") There might be *bridge* authorities that exist only to cross-certify, or just a loosely organized *oligarchy* of independent roots. All of these approaches have emerged in current PKIs, with varying degrees of implementation and engineering complexity—it's been hard to get it right.

Trust paths

An artifact of moving from a single universal root to a more complex network is that it complicates the notion of *trust path*: the route from a relying party's root to a target certificate. In the simple model, the trust path *is* the certificate: the One True Root signed this, so we believe it. In a more complex model, we need to figure out how to construct the trust path, and what the semantics mean (for example, consider the composition of two cross-certificates described in the previous section). When a relying party *P* needs to make a judgment about certificate *C*, whose trust path may be long, we also need to figure out how to get all the other certificates *P* might need to *P*. Suddenly we might need directories and repositories, and the space in protocols and handshakes and tables we implicitly assumed would hold one certificate may now hold several. (Indeed, current PKI-based tools can break when an extra certificate gets introduced into paths.)

How we make this scale to a population the size of the IoT is not trivial.

Revocation

Secrets become nonsecret. In the current world, people lose credit cards and college/employer ID cards; people divulge passwords; activist hackers (and presumably more secretive malicious ones) penetrate systems to obtain keys (even high-value private keys). Even in these cases, human individuals perceive a motivation to keep their credit cards or login dongles close at hand.

In the envisioned IoT, we will have far more devices distributed in more uncertain environments. (Who exactly has access to that box? Will they have motivation to protect it—or to compromise it?) Furthermore, the state of the art in having physical devices protect their own secrets is a continual game of cat-and-mouse between attack and defense technology (e.g., [17, 20]); if grid devices are to be affordable and long-lived, it's probably safer to assume that adversaries will be able to extract secrets if they get destructive physical access.

Consequently, a PKI needs to allow for the fact that any given certificate may need to be suddenly *revoked*: "Oops, it's no longer true that the thing that knows the private key matching the public E_x is necessarily *X*." The necessity of potential revocation gives rise to a new problem: how does a relying party know if a given certificate has been revoked? (If nontrivial trust root structure has given us nontrivial trust paths, the problem is compounded: the relying party needs to do this for each certificate in a potential trust path.)

The traditional PKI approach to this problem has been for a CA to regularly publish a *certificate revocation list* (CRL). In theory, a relying party regularly obtains a fresh CRL; it assumes this is valid until the next CRL is published, and in the meantime checks if each new target certificate is present in the list. In practice, this hasn't worked too well. Johnson & Johnson initially set up a PKI for 60,000 employees, and ended up with a CRL an order of magnitude larger than anticipated, due to a user interface quirk [7]; other domains (e.g., [13, 10]) also have seen CRLs significantly larger than expected, straining bandwidth and exacerbating latencies. In the web's SSL, revocation is widely regarded as not actually working.

Researchers have explored alternatives, such as the *online certificate status protocol* (OCSP), where the relying party checks a certificate's revocation status with a trusted entity in real time, or hash-chained schemes (e.g., [11]) to reduce the bandwidth for revocation data. Nonetheless, even in current large-scale PKIs, revocation is a challenge. It doesn't scale.

IoT scale

To reiterate:

- Johnson & Johnson's PKI initially had only 60,000 certificates, but faced scalability challenges with revocation.
- The web's SSL has (according to some surveys) only a few million properly certified keyholders, and revocation doesn't work.

How is a PKI for the billions of things in the IoT going to fare better?

CONSTRAINED DEVICES AND CHANNELS

Cryptographic computation can be expensive, in terms of both time and space. For example, calculating an RSA digital signature may require raising a 2,048-bit integer to the power of another 2,048-bit integer—intensive for a CPU whose native word size may be only 32 bits. And if the item being signed is only a few bytes, an RSA signature that may take 2,048 bits of storage represents a rather large overhead (although newer elliptic curve techniques can reduce this overhead somewhat). The computation costs arise when we think about devices; the data size costs arise when we think about device memory and communication channels, where slow channels may turn data cost into additional time cost.

These costs have already proven problematic in the incipient IoT. Embedded systems often use previous-generation processors, and avoid expensive cryptography as a result. Devices with limited electric power also need to consider computational cost: for example, it has been observed that the "obvious" solution to securing the wireless interface on an embedded cardiac device—adding strong cryptographic authentication—enables a new way of killing the patient, by repeated attempting cryptographically flawed authentication and draining the battery.

The current power grid still makes extensive use of slow serial networks, where the additional time necessary to collect the bits for a digital signature on a control message—even before verification—delays the message unacceptably. In my own lab's prior work on BGP, which controls the routing behavior of the IoC, we found that adding standard cryptographic authentication (and many variations) killed performance, both in terms of the time it takes for the system to reachieve stability after a significant routing change and in message and caching costs (e.g., [21]).

If we project to the envisioned IoT, it's natural to expect more of these constrained devices and channels, and the concomitant engineering challenges for cryptography. One response to these constraints is to focus on more lightweight schemes, such as Google's macaroons (as already mentioned) or the *Simon* and *Speck* symmetric ciphers, designed by researchers at the NSA and tuned for limited-power devices [2]. Over the last 15 years, there has also been a lot of attention paid to elliptic curve cryptography (as noted previously), a branch of public key cryptography that get can get more security with smaller keys and signatures and such.

Another response is to offload complex cryptographic work from the IoT device itself to an *avatar* somewhere in the cloud. The device then merely needs to maintain a secure connection to its avatar, which does all the heavy lifting. Zebra Technologies explicitly uses the "avatar" term in this context, although it's essentially the approach taken by other early IoT deployments, such as NEST.

PRIVACY SIDE EFFECTS

A cryptographic protocol to do something like prove identity is usually intended to do just that—at the end of the dance between Bob and Alice, Bob has some kind of evidence that the party at the other end is Alice. However, in straightforward implementations, Bob may learn a whole lot more. For example, if Alice is using standard X.509 PKI, then Bob learns everything in her certificate, as well as who issued it, and what the issuer's certificate looks like, and so on up the trust chain. The trust path supporting the certificate of a smart appliance X may betray the path of ownership, landlord, and subletting; the revocation or directory queries a substation server *S* receives from neighborhood consumer devices can betray their activity.

To use a current real-world example, suppose all cash transactions were suddenly replaced with credit cards. Purchasing a book with physical cash suffices for the buyer Alice to authenticate to the seller Bob that she has sufficient funds; but if Alice uses a credit card, then a far more extensive data trail emerges. If this larger data spill is a concern, cryptographic means exist (at least in the research world) to limit Bob from learning extra information about Alice—and even from learning anything except that she is authorized. However, this dimension of lost anonymity needs to be considered and addressed before the IoT is built, not afterward.

CRYPTOGRAPHIC DECAY

The mechanisms for identification and authentication in the IoT will likely depend on cryptographic techniques, as they did in the IoC. At this book has repeated, the IoT will differ from the IoC in many aspects, including these two:

Lifetime
: IoT things, such as household appliances or sensors embedded in urban infrastructure, will likely live much longer than traditional IoC computers.

Patching
: IoT things may be much harder to reach with software updates—and the vendors responsible for developing and pushing those updates may no longer exist.

As a result, the software an IoT thing has when it ships may be the software it has for decades. This means the cryptographic algorithms and their relevant parameters (such as key length) could remained fixed for decades.

This is of concern for several reasons. To begin with, as this chapter noted earlier, practical cryptography usually has unsatisfyingly soft impossibility thresholds. To discover a private or secret key, brute-force searches are always potentially possible, and in the best case for defenders, they are limited only by the current level of computational power—which historically tends to increase exponentially. For example, if an adversary can factor the public modulus for an RSA keypair, he or she can quickly calculate the private key. RSA modulus lengths of 1,024 to 2,048 bits are considered secure today, but are projected to be breakable by 2030. Should devices that ship today become by default insecure by 2030?

Cryptographic analysts try to make well-founded predictions today about key length lifetime, but it's important to note that, historically, the field has often overestimated security against brute-force. For example, the original RSA paper proposed modulus lengths of less than 700 bits—something considered completely insecure now.

Of even more concern than the gradual decay from Moore's Law is the potential of sudden decay due to new insights into algorithms or computation. Looking forward, it has already been shown that if quantum computers (see, section 8.5 in my security textbook [15]). can be built, then factoring large numbers will become efficient. A sudden breakthrough in this engineering problem will break the cryptography baked into the IoT—and even without quantum, the possibility still exists that a clever theoretician somewhere will show that factoring is efficient even on traditional computers.

Looking backward, we see that sudden breaks like this have already happened:

- In a few short years after the turn of the century, the MD5 hash algorithm went from being considered probably secure, to having semantically meaningless collisions creatable with prohibitively large amounts of computing power, to having semantically meaningful collisions creatable in negligible time on a laptop.
- In the same period, the ISO9796-1 algorithm used to extend hash values to full modulus length went from being a best practice for secure digital signatures to completely breakable.

Of course, this discussion does not even into take into account the fact that IoT cryptography, like any other software, may suffer from bugs. IoC cryptography has already seen fatal implementation blunders, such as predictable numbers being used for critical values such as keys, or private keys being discoverable from observables such as computation time—and embedded system cryptography has already been shown vulnerable to attacks via power measurement or carefully chosen interruption of power.

In a nutshell, expecting IoT devices to remain secure if they remain baked in with cryptography considered secure at shipping time does not bode well. Potential countermeasures include ensuring that all devices can be patched quickly, or can easily die if they persist beyond their secure lifetime, or can be shipped with aggressively future-proof cryptography. Again, challenges remain.

Moving Forward

These will be interesting times. Will we end up with a global IoT PKI that goes where no PKI has gone before? Will IPv6 bring a secure DNSSEC infrastructure that enables universal IoT names and attributes? Or will some particular legacy application gain a sufficiently strong footing that all IoT naming and authentication will bootstrap off of that? One might facetiously imagine an extension of Facebook called Thingbook, where each smart thing has a profile and sends messages and wall posts to its peers, all managed by other Thingbook things—and then realize that perhaps rather than joking, one should quickly file a patent application, because maybe that's what might do the trick.[1]

Works Cited

1. P. Anantharaman, K. Palani, D. Nicol, S. W. Smith, "I am Joe's fridge: Scalable identity in the Internet of Things," in *Proceedings of the 9th International Conference on the Internet of Things*, December 2016.
2. R. Beaulieu and others, "The SIMON and SPECK lightweight block ciphers," in *Proceedings of the ACM Annual Design Automation Conference*, 2015.
3. A. Birgisson and others, "Macaroons: Cookies with contextual caveats for decentralized authorization in the cloud," in *Proceedings of the Network and Distributed System Security Symposium*, 2014.
4. D. Chadwick, A. Otenko, and E. Ball, "Role-based access control with X. 509 attribute certificates," *IEEE Internet Computing*, March/April 2003.
5. D. Clark, J. Elien and others, "Certificate chain discovery in SPKI/SDSI," *Journal of Computer Security*, 2001.
6. B. Gassend and others, "Silicon physical random functions," in *Proceedings of the ACM Conference on Computer and Communications Security*, 2002.

1 Parts of this chapter are adapted from my papers [1] and [16].

7. R. Guida and others, "Deploying and using public key technology: Lessons learned in real life," *IEEE Security and Privacy*, July/August 2004.
8. P. Gutmann, "PKI: It's not dead, just resting," *IEEE Computer*, August 2002.
9. R. Housley and T. Polk, *Planning for PKI*. John Wiley and Sons, 2001.
10. P. Kehrer, "Defective by design? Certificate revocation behavior in modern browsers," *SpiderLabs Blog*, April 4, 2011.
11. S. Micali, "NOVOMODO: Scalable certificate validation and simplified PKI management," in *Proceedings of the 1st Annual PKI Research Workshop*, 2003.
12. S. Nakamoto, "Bitcoin: A peer-to-peer electronic cash system," *Bitcoin*, October, 2008.
13. R. Nielsen, "Observations from the deployment of a large scale PKI," in *Proceedings of the 4th Annual PKI R&D Workshop*, 2005.
14. R. Rivest and B. Lampson, "SDSI—A simple distributed security infrastructure," April 1996.
15. S. Smith and J. Marchesini, *The Craft of System Security*. Addison-Wesley, 2008.
16. S. W. Smith, "Cryptographic Scalability Challenges in the Smart Grid," in IEEE PES Innovative Smart Grid Technologies, January 2012.
17. S. W. Smith, "Fairy dust, secrets, and the real world," *IEEE Security and Privacy*, January/February 2003.
18. F. Stajano and R. Anderson, "The resurrecting duckling: Security issues in ad-hoc wireless networks," in *Proceedings of the 7th International Workshop on Security Protocols*, 1999.
19. R. P. Timothy, J. Pierson Xiaohui Liang, and D. Kotz, "Wanda: Securely introducing mobile devices," in *Proceedings of the IEEE International Conference on Computer Communications (INFOCOM)*, April 2016.
20. S. Weingart, "Physical security devices for computer subsystems: A survey of attacks and defenses," in *Cryptographic Hardware and Embedded Systems —CHES 2000*, 2000.

21. M. Zhao, S. W. Smith, and D. M. Nicol, "Aggregated path authentication for efficient BGP security," in *Proceedings of the 12th ACM Conference on Computer and Communications Security*, 2005.

The Internet of Tattletale Devices

The IoT's deep reach into society means we will need to reexamine how we think about technology and privacy.

In the emerging IoT, the "things" connect more deeply into an individual's life and behavior than did the computers of previous IT infrastructure. They also transmit this data to a wider range of players. Consequently, the IoT may enable aggregation across large populations of individuals and across larger populations of attributes for any one individual.

These new dimensions challenge the way we think about *privacy*: access and control of data about ourselves. This chapter considers some of these challenges, discussing:

- Cautionary tales about IoT and IoC privacy leakage
- Areas where current IoT collection of data was intended but surprising
- How the emerging IoT may enable widespread monitoring of individuals
- Why an effective privacy policy is a hard goal to achieve

Cautionary Tales

Chapter 2 discussed how, in a typical IoT architecture, distributed sensors and actuators connect to a big data backend, and Chapter 3 observed how the future has been here before. Let's take a look at a few cautionary tales.

IOC PRIVACY SPILLS

Big data backends already have a history of privacy spills. Spills of personal data records have become so common that they're barely newsworthy anymore. One nicely ironic incident was the 2015 compromise of the US government's Office of Personnel Management (OPM); information (including mine) collected during background investigations of government employees, in order to ensure their trustworthiness, was then provided in bulk to attackers apparently working for a foreign nation [18]. In the medical domain alone, *Forbes* reports over 112 million records were spilled in 2015 [43].

Many of these spills are due to attackers breaching the servers where they are stored—a standard matter of a hole or lack of patching. In 2015, BinaryEdge surveyed the internet for installations of four specific database tools and found over a petabyte (10^{15} bytes) of data exposed online due to lack of authentication [3] (recall Chapter 4). Some leaks stem from lost laptops, but others arise from the complexities of connection. For one example, the FTC reports [47]:

> *Medical transcript files prepared between March 2011 and October 2011 by Fedtrans, GMR's service provider, were indexed by a major internet search engine and were publicly available to anyone using the search engine. Some of the files contained notes from medical examinations of children and other highly sensitive medical information, such as information about psychiatric disorders, alcohol use, drug abuse, and pregnancy loss.*

For another example, the White House bragged about moving citizen services from paper to electronic [58]:

> *In 2014, the Internal Revenue Service made it possible for tax-payers to digitally access their last three years of tax information through a tool called Get Transcript. Individual taxpayers can use Get Transcript to download a record of past tax returns, which makes it easier to apply for mortgages, student loans, and business loans, or to prepare future tax filings.*

Unfortunately, in 2015, attackers used this "Get Transcript" service to download other people's past returns, and then, using this information, filed fraudulent returns for the current year and collected the tax overpayments. The *New York Times* reports that over $50 million was stolen from over 100,000 individuals [53].

Given that the IoT will have a larger number of systems interconnected more complexly, we also might expect more spills resulting from inadvertent interconnection. What else do these IoC issues say about privacy in the IoT?

For one thing, as we go from the IoC to the IoT, there's no reason to suspect the backend servers will become *more* secure. It's also interesting to note that root factors in the OPM case included dependence on software platforms no longer supported by their vendors, and on web portals using obsolete cryptographic protocols. As Chapter 1 noted, the IoT will likely bring lifetime mismatches among things, software, and vendors; as Chapter 4 noted, the lifetime of things (and the difficulty of patching them) may lead to more trouble with aging cryptography.

We also might expect the IoT to inherit the risk of exposure from physical distribution, since it will have even more separate pieces more broadly distributed, and more likely to be misplaced. Indeed, things may even live longer than the people and enterprises that use them—and researchers and journalists already report finding interesting personal data on used machines purchased on eBay and elsewhere. Even now in the IoC, a problem here is the deep ways that data can penetrate and hide in systems. A colleague who worked in IT at a large hospital used to report that when clinicians would return borrowed laptops—after conscientiously trying to scrub them of patient data—he could always find some such data remaining somewhere. Tools that would help purge devices of such sensitive remnants would help both in the IoC and the IoT.

Unfortunately, we are already seeing risks to the IoT from IoC privacy problems becoming more than just theoretical. In 2015, *Motherboard* reported on a spill of backend data from VTech, an IT-enhanced toy company [16]:

> *The personal information of almost 5 million parents and more than 200,000 kids was exposed earlier this month after a hacker broke into the servers of a Chinese company that sells kids toys and gadgets.... The hacked data includes names, email addresses, passwords, and home addresses.... The dump also includes the first names, genders and birthdays of more than 200,000 kids. What's worse, it's possible to link the children to their parents, exposing the kids' full identities and where they live*

IOT PRIVACY WORRIES

Besides merely inheriting the privacy risks of their backend servers, IoT products and applications have also been introducing new concerns of their own.

Adding "smart" functionality to previously less-smart home appliances is one area where this concern is manifested. For example, consider Vizio smart TVs [31]:

Vizio has sold more than 15 million smart TVs, with about 61 percent of them connected as of the end of June [2015]. While viewers are benefiting from those connections, streaming over 3 billion hours of content, Vizio says it's watching them too, with Inscape software embedded in the screens that can track anything you're playing on it—even if it's from cable TV, videogame systems and streaming devices.

Of additional concern is the 2015 discovery that, thanks to the "bad PKI" pattern described in Chapter 4, adversaries can intercept the Vizio data transmissions [22]. The TV, its corporate partners, and adversarial middlemen are watching you. Robert Bork[1] might be rolling over in his grave.

Adding voice interaction to home devices adds another vector of concern. Apple plans to have Siri listen to and transcribe her owner's voicemail, using various servers and services in the process [11]. Samsung smart TVs have a voice recognition feature that allows users to control the TV by speaking to it. Implementing this feature requires sending the audio back through the cloud, including to players other than Samsung. *ITworld* notes this advice from Samsung [10]:

Please be aware that if your spoken words include personal or other sensitive information, that information will be among the data captured and transmitted to a third party through your use of Voice Recognition.

Your TV is also listening to your living room!

It's interesting to note the Samsung implementation also followed the insecurity design pattern of forgetting to encrypt [9]:

Following the incident, David Lodge, a researcher with a U.K.-based security firm called Pen Test Partners, intercepted and analyzed the Internet

1 History lesson: when Robert Bork was nominated for the U.S. Supreme Court in 1987, a reporter obtained and published his video rental records—which led to Congress passing a law making this particular kind of privacy spill illegal.

> *traffic generated by a Samsung smart TV and found that it does send captured voice data to a remote server using a connection on port 443.*
>
> *This port is typically associated with encrypted HTTPS (HTTP with SSL, or Secure Sockets Layer) communications, but when Lodge looked at the actual traffic he was surprised to see that it wasn't actually encrypted.*

(As we move forward into the IoT, "closing the loop" with testing like this will be vital.)

In 2015, Mattel announced "Hello Barbie," bringing cloud-connected voice interaction to children's dolls [55]. The *Register* described it as:

> *A high-tech Barbie that will listen to your child, record its words, send them over the internet for processing, and talk back to your kid. It will email you, as a parent, highlights of your youngster's conversations with the toy.*

The article further noted the conversations might persist in the cloud, as grist for the computational engines. According to ToyTalk's privacy policy:

> *When users interact with ToyTalk, we may capture photographs or audio or video recordings (the 'Recordings') of such interactions, depending upon the particular application being used. We may use, transcribe and store such Recordings to provide and maintain the Service.*

Although "Hello Barbie" repeated some cryptographic insecurity patterns (as Chapter 4 discussed), no major privacy disaster has emerged. However, the implications are deep. Is one's home still a sanctuary? Who gets to know what one's child whispers to a toy? Will these recordings result in parents being visited by child welfare agents—or (depending on the society) being prosecuted for political thoughtcrime? Will the whispers show up when the child grows and applies for a job or a security clearance, or is a defendant in court?

The consumer-side smart grid—a sexy IoT application domain discussed back in Chapter 3—has also triggered privacy concerns. In its intended functionality, the smart grid will instrument appliances and other electrical devices in the home with sensors and maybe even actuators so that the rest of the grid can better keep things in balance (bringing the user along by offering better pricing and such). The privacy concern here is that this set of measurements—how much power each device in a house or apartment is using at any given moment in time —becomes a signature of what is going on there: who is present and what they are doing. The patterns that emerge over time can then be the basis for prediction of future household activity—useful to the grid, perhaps, but also useful to criminals who might want to know when the home is empty, or when an intended assault or kidnap victim is there alone.

As with any privacy exposure in information technology, the issue exists on two levels: not only what the authorized backend entities might know, but also what might be gleaned by unauthorized entities who penetrate communication channels or servers.

When Things Betray Their Owners

The preceding scenarios were mostly describing *potential* privacy problems in the IoT, and were inspired by IoC privacy issues. However, there have already been many *actual* privacy problems arising from what IoT things actually do: collect lots of data about the physical world around them—including their owners.

YOUR THINGS MAY TALK TO POLICE

In March 2015 in Pennsylvania, a woman called 911 to report she had been raped by someone who broke into her house and assaulted her while she was sleeping. However, police investigators concluded she had made the story up, in part because of her Fitbit [24]:

> *The device, which monitors a person's activity and sleep, showed Risley was awake and walking around at the time she claimed she was sleeping.*

Big Brother may not be watching you, but Little Fitbit is.

In December 2015 in Florida, a driver (allegedly) was involved in a car accident and then fled [40]. When telephoned by an emergency dispatcher, she responded, "Ma'am, there's no problem. Everything was fine." Hit-and-run accidents are nothing new. However, what's interesting in this case is *how* the police became involved:

> *The dispatcher responds: "OK, but* your car called in *saying you'd been involved in an accident. It doesn't do that for no reason. Did you leave the scene of an accident?" [Emphasis added]*

The woman's car, like many new cars enhanced with computing magic, was set up to call 911 if its GPS information indicated a potential crash (e.g., because of a rapid change in direction and momentum). Your car can know where you are and will call 911 to help you, even if that's not your plan. But what else will happen with this data? The article notes further:

> *Privacy campaigners concerned that governments might use the technology to keep permanent track of a vehicle's movements have been told the new rules only allow for GPS information to be collected in the event of a collision, and that it must be deleted one it's been used.*

But will the GPS data really "only" be used for the noble purpose of accident reporting? Indeed, modern IT-enhanced cars collect much data, and police know about this.

In Vermont in 2015, a cyclist was killed by a car whose driver was allegedly intoxicated [13]:

> *Scott said investigators obtained a search warrant for the Gonyeau car to download information from its computer. He said once the information from the car's sensors can be reviewed, police will know more about the crash.*

The investigation later concluded the cyclist was mostly to blame.

In 2016, Canada's CBC News reported [6]:

> *From July 1 to Dec. 31 of last year, there were five fatal vehicle collisions in the parts of Halifax policed by the RCMP. Information from event data recorders was used in two of those investigations, according to an access to information request filed by CBC News.*

CBC also noted the various implications:

- Are the owners aware their cars collect this data?
- Will the data only speak of things such as car accidents, and not other aspects of driver and passenger identity and behavior?
- Will the police only use the data for correct purposes?

- Is the data actually correct?

It's worth noting that a colleague of mine who spent a career in law enforcement (in the US, a country with constitutional privacy protections for citizens) observed that it's common practice for police to use illegal means to find out who's guilty—after which they then use legal means to obtain evidence for court. It's also worth noting that just because a computer allegedly measured something doesn't mean that it actually happened; "Things 'on the witness stand'" on page 180 will consider further the legal implications.

YOUR THINGS MAY PHONE HOME

Law enforcement officials aren't the only people your smart things may talk to.

In February 2013, John Broder wrote an unfavorable review of the Tesla Model S in the *New York Times* [5]. Broder was unhappy with the performance of the high-end electric car, and supported this conclusion with his firsthand observations of speeds and charges and such as he test-drove it. What's interesting here from the IoT perspective is that the reviewer was not the only witness—the car itself was recording data about its experiences and sending this data back to Tesla. Unhappy with the review, Tesla chair Elon Musk published a retort [44] using the car's logs to dispute what the reviewer claimed happened during this "most peculiar test drive." For example, one of the diagrams showed a speed versus distance graph, with annotations appearing to show that Broder's claims of "cruise control set at 54 miles per hour" and how he "limped along at about 45 mph" did not match recorded reality. A back-and-forth ensued, with no clear winner [23]. (Tesla would not give me permission to republish any of these diagrams, but you can see them in [44, on page T3].)

In 2016, this pattern continues, with high-profile incidents (e.g., [20]) of customers claiming their Teslas did something odd, and Tesla using its logs to claim otherwise.

In the IoT, your things are also witnesses to what you witness—and they may see it differently.

Given the computationally intensive engineering challenges of high-tech and high-end cars such as Teslas, the fact that they log data and send it back home would appear reasonable. The more data is collected, the more the engineers can analyze and tune both the design in general and that car in particular. Tesla is not alone in doing this. One colleague reported his BMW decided it needed servicing and told BMW, which called my colleague—while he was driving. (The message was something like "Your car says it needs to be serviced right now.") Another colleague who handles IoT security for the company whose machines generate "half the world's electricity" talks about the incredible utility of being able to instrument complex machines, send the data back home, and feed it into computerized models that the engineers trust more than physical observation.

However, in February 2015, Brian Krebs wrote about a family of IoT devices that appear to phone home for no reason at all [30]:

> *Imagine buying an internet-enabled surveillance camera...only to find that it secretly and constantly phones home to a vast peer-to-peer (P2P) network run by the Chinese manufacturer of the hardware.*

In fact, this is what newer IP cameras from Foscam were doing—which came to light when a user "noticed his IP camera was noisily and incessantly calling out to more than a dozen online hosts in almost as many countries." To make things even more interesting, the camera UI does allow the user to tick a box opting out of P2P—but doesn't actually change its behavior when the box is ticked. In this case, it's harder to see a reasonable argument for the P2P network; Foscam claims it helps with maintenance.

YOUR THINGS MAY TALK TO THE WRONG PEOPLE

In the preceding cases, IoT things shared data about their experiences in perhaps surprising ways—but at least they were sharing it in accordance with their design (e.g., to authorized law enforcement officers, or to the original vendor for maintenance and tuning).

However, a problem with exposing interfaces is that, perhaps due to one of the standard insecurity patterns of Chapter 4 or perhaps due to a new one, one may inadvertently provide these services to more parties than one intended. Unfortunately, this has already happened with IoT data collection.

GM brags that its OnStar system for collecting and transmitting car data has "been the Kleenex of telematics for a long time" [19]. In 2011, Volt owner Mike Rosack so much enjoyed tracking the telematics he received on his phone from his car that he reverse-engineered the protocol and set up the Volt Stats website, which enabled a broader population of Volt owners to share their telematics. Unfortunately, doing this required that the owners share their credentials with Volt Stats (the "lack of delegation" pattern from Chapter 4). GM decided this was an unacceptable privacy risk and shut down the API, but then provided an alternate one that allowed Volt Stats data sharing to continue but without this risk.

Unfortunately, in 2015, researcher Samy Kamkar found a way to surreptitiously capture owner credentials (the "easy exposure" pattern from Chapter 4). The resulting OwnStar tool allows unauthorized adversaries to usurp all owner rights [17].

In Australia, four shopping malls set up "smart parking" that used license-plate readers to track when cars entered and left, and gave users the option of receiving text alerts when their parking time was close to expiring. However, the malls discontinued this service when it was noticed that anyone could request notification for any vehicle (the "no authentication" pattern from Chapter 4) [12].

Chapter 1 discussed the Waze crowdsourcing traffic mapping application. Chapter 4 mentioned the "bad PKI" design pattern that has been surfacing in IoT applications. One place it has surfaced is in Waze: in 2016, scholars at UC Santa Barbara demonstrated that (due to flaws in checking certificates) they could intercept Waze's encrypted SSL communications, and then introduce "thousands of 'ghost drivers' that can monitor the drivers around them—an exploit that could be used to track Waze users in real-time" [26]. Here, the service being usurped by the unauthorized party ("Where is driver X right now?") was not really one of the intended services to begin with.

As an extreme case of unauthorized access to unintended services, researchers at SRI (recall Figure 1-3 in Chapter 1) have been worrying about not just adversarial access to the internal IT of government automobiles, but even mere detection that a particular vehicle is passing by. For a terrorist or assassin, the ability to build a roadside bomb that explodes when one particular vehicle goes by would be useful indeed. In this case, even the natural solution of "disable all electronic emanations" would not work, since the bomb could simply wait for the car that is suspiciously silent.

Emerging Infrastructure for Spying

The previous section closed with thinking about how the IoT could be useful for terrorists. IoT applications can also have utility for adversaries (such as spies or corrupt government officials) interested in systematically monitoring an individual's activity.

WEARABLES AND HEALTH

One set of issues arises from IoT technology tied to a person: smartphones and applications, Fitbits and other wearables, Garmin-style devices on bicycles, mobile health (mHealth) technologies, etc.

Such technology can have upsides for the individual: monitoring aspects of health, tuning and improving athletics, tracking and sharing bicycle and running routes, etc. Friends concerned about their weight track calorie intake with iPhone applications connected to databases of food items; friends interested in exploring fine beer track their drinking with a similar application. When I was a serious bike racer, everyone in the peloton started using wearable heart-rate monitors to track and tune performance. The local cycling and running communities are full of people who religiously monitor each ride or run with Strava.

Initial privacy (and security) concerns about wearables arise from straightforward issues:

- By design, can they expose data to the wrong parties? For a positive example, Strava addresses this concern by permitting users to set up a *privacy zone* around their residence, so that cycling or running routes from there will actually appear to have an endpoint somewhere nearby.
- Does the core device have secure interfaces? For a negative example, in 2015 researcher Simone Margaritelli discovered that the Nike+ Fuelband fell into some of the standard patterns from Chapter 4: flawed authentication allowed "anyone to connect to your device," and inadvertent inclusion of debug code allowed the ability to alter the internal programming [34].
- Does the device's supporting infrastructure have problems? For example, in a survey of Android mHealth applications, researchers from the University of Illinois at Urbana-Champaign discovered many did not encrypt communications (thus exposing user data to anyone listening in) and used third-party services (thus incurring privacy dependence on cloud parties the user might not know about) [25].

Beyond these straightforward issues, deeper privacy issues arise when one considers who *else* (besides the user) might benefit from these personal technologies.

The fact that many (US) employers also pay for their employees' health insurance costs creates a business motivation for employers to promote wellness via Fitbits or the like, as Chapter 7 will discuss. However, what else should the employer know? In California in 2015, a woman filed a lawsuit claiming her employer required her to run an iPhone application that "that tracked her every move 24 hours a day, seven days a week" [29]. In 2016, the Dutch government ruled that employer monitoring of employees via wearables violated employee privacy, even if the employee gave consent, since the power relationship makes true consent impossible [46].

What if the business itself depends on physical performance? At a wearables event in Canada connected to the 2015 Pan Am games, panelists discussed a variety of issues, ranging from upsides for everyone (better performance) to the privacy downsides for athletes [59]:

> *If predictive analytics can suggest when an athlete's performance will start to decline, will team owners use that data to shorten players' contracts or pay them less, even while they're still at the top of their game?*

Yet another issue: who actually owns the athlete's data? (In the US, we've already seen a similar kerfuffle for the general population: who owns your health record?)

The provider of a wearable service benefits monetarily, and advertisers benefit by gaining precise information about their target audience. Putting the two together, a wearable service might expose a user's personal information to advertisers. In fact, in May 2016, new reports indicated the operators of Runkeeper were doing that, and more, in apparent violation of European privacy laws: "It turns out that Runkeeper tracks its users' location all the time—not just when the app is active—and sends that data to advertisers" [7].

INTERNET OF BIG BROTHER'S THINGS

George Orwell's *1984* posited a dystopian future where a governmental Big Brother monitored every move of every citizen. Many aspects of the IoT may be useful to Big Brothers (government or otherwise).

To start with, many current IoT applications seem targeted directly to improving surveillance. In 2015, the *Mercury News* reported that the city of San Jose added license-plate readers to trash trucks so that they could scan and report to the police what cars they saw along their routes [21]. The ACLU voiced objections:

> *If it's collected repeatedly over a long period of time, it can reveal intimate data about you like attending a religious service or a gay bar. People have a right to live their lives without constantly being monitored by the government.*

For other examples of fusion, Bill Schrier, formerly the CTO of Seattle, speculates on the "Internet of First Responder Things" [51]; Macon-Bibb County in Georgia is considering deploying drones as first responders [14].

Big Brother can also watch via cameras. While CCTV cameras in urban areas have been emerging over the last few decades, recent years have seen commercial products that embed such things in other devices. Sensity (*http://www.sensity.com/for-airports-1/*) bundles active surveillance into previously inert infrastructure such as lighting:

> *NetSense for Airports offers a unique and effective solution for enhanced security through the airport terminal, parking lots and perimeter. By embedding video and other sensors inside LED luminaires, energy savings are combined with high-power security technology.*

The ACLU again expressed concern [36]:

> *These lightbulbs-of-the-not-so-distant-future will also be able to GPS track individual shoppers as they travel through stores. Wait. What? The light bulbs can function as tracking devices? We would have to imagine that if they can GPS-track shoppers in stores, they could work just as effectively to track people as they walk the streets of our cities and towns. In fact, if you traveled through Newark Liberty International Airport in the past year, these spy-bulbs lights were already watching you. And there's more: the bulbs can be programmed to "pick() up on suspicious behavior." What exactly does that mean? If two women wearing head scarves decide to chat in a parking lot after seeing a late night movie, are the police going to be notified?*

On a lighter note, mass video surveillance has an upside: the band The Get Out Clause allegedly used ubiquitous government cameras to film their breakout music video [8]:

> *"We wanted to produce something that looked good and that wasn't too expensive to do," guitarist Tony Churnside told Sky News.*
>
> *"We hit upon the idea of going into Manchester and setting up in front of cameras we knew would be filming and then requesting that footage under the Freedom Of Information act."*

One can harvest entertaining screenshots from this video; however, I could not find anyone to ask for publication permission, so no screenshot will appear here.

In 2016, Matt Novak of Paleofuture observed [45]:

> *Back in March, I filed a Freedom of Information request with the FBI asking if the agency had ever wiretapped an Amazon Echo. This week I got a response: "We can neither confirm nor deny."*

In 2016, the *Guardian* reported that James Clapper, former US Director of National Intelligence, had observed [1]:

> *"In the future, intelligence services might use the [Internet of Things] for identification, surveillance, monitoring, location tracking, and targeting for recruitment, or to gain access to networks or user credentials."*

Richard Ledgett of the NSA concurred [37]:

> *Biomedical devices could be..."a tool in the toolbox".... When asked if the entire scope of the Internet of Things—billions of interconnected devices—would be "a security nightmare or a signals intelligence bonanza," he replied, "Both."*

In 2016, the US has seen ongoing debate about whether law enforcement officials should have warrantless access to citizens' prescription information [41]:

> *"It has become the status quo that when a person comes under their radar they run to the prescription drug database and see what they are taking," said Sen. Todd Weiler, a Republican—who said that police in Utah searched the PDMP database as many as 11,000 times in one year alone. "If a police officer showed up at your home and wanted to look in your medicine cabinet and you said no, he would have to go and get a search warrant."*

Interestingly, such access has actually *enabled* drug abuse, as witnessed by the report of:

> *An opioid addicted police officer who was caught on video stealing pills from an elderly couple's home after tracking their prescriptions in the state's PDMP database.*

On a brighter note, in 2014 the Supreme Court ruled, in *Riley v. California*, against warrantless searching of a cellphone [32]:

> *Even the word cellphone is a misnomer, [Chief Justice Roberts] said. "They could just as easily be called cameras, video players, Rolodexes, calendars, tape recorders, libraries, diaries, albums, televisions, maps or newspapers."*

Instead of asking what's possible, perhaps we should be asking what isn't possible. Researchers from Nanjing University have shown that accelerometers alone (and not GPS or cellular connection) can suffice to track an individual's motion through an underground train system [48]. Researchers here at Dartmouth College have shown data from a student's smartphone can predict both GPA and psychological depression [39]. As I wrote these words, news reports indicated that (thanks to the "default password" pattern) the Quebec Liberal Party's videoconference system could be used to spy on party meetings [15]. No wonder *InfoWorld*'s Fahmida Rashid fears the Google Home of the future [50]:

> *Always-listening devices accelerate our transformation into a constantly surveilled society. That's a problem not only for us but for our kids, too.*

Getting What We Want

Current and proposed IoT applications have privacy risks. What can we do about this?

SAYING WHAT WE WANT

First, there's an old saying in software engineering that if you can't say what correct behavior is, then the system can never behave incorrectly. An overarching problem with saying the IoT will reveal too much personal information to the wrong parties is what such a statement implicitly implies:

- That some amount of information is "not too much"
- That some parties are acceptable recipients of this information
- That we (as individuals or as a society) are able to express exactly what these rules are
- That computers are able to enforce these rules

Let the individual decide?

For example, in the case of electronic embodiments of personal health information, one often hears the assertion that each individual patient should be able to decide what to share with whom. Although perhaps compelling, this doctrine has problems.

First, it's not what happens in the pre-IoT health world. In the US, law requires that clinicians treating certain kinds of issues (e.g., gunshot wounds, suicidal tendencies, potential molestation, certain communicable diseases) report these to various authorities. The question of whether a parent can legally see health information about his or her child is also surprisingly nuanced.

Secondly, the patient may not necessarily be in the position to judge "need to know" accurately. An MD friend who works in medical informatics loves to point out annoying counterexamples:

- In the 1984 Libby Zion case, an 18-year-old woman receiving emergency treatment in a New York City hospital died unexpectedly, leading to lengthy litigation. According to some versions of the story, Zion died in part because neither she nor her family decided it was relevant to tell clinicians she had been on antidepressants and had recently taken cocaine. (See [28] for one viewpoint.)
- Should your heart doctor know about your dental health? Current consensus in the medical community says yes: gum issues connect to heart disease [2].

Another challenge is that the problem of Alice deciding to share *X* with Bob is not well defined. When does Alice make this decision? When she makes the decision, does she express what she really wants? Over the last few decades, psychology has produced many reproducible results in *cognitive bias*: how human minds can form perceptions and judgments in surprisingly bizarre and "incorrect" ways. In my own research, I've looked at how cognitive bias can complicate security and privacy [54]. For example:

- Perhaps due to *dual process* issues, educating users about social network privacy issues can lead them to make quantitatively worse privacy decisions [56].
- Perhaps due to the *empathy gap*, reasonable electronic medical record (EMR) users in policy meetings will make access control decisions that reasonable EMR users in practice will find overly constraining [57].

An older friend of mine in the US even laments that he would like to have all of his medical information widely accessible by default, since that might lead to better treatment as he ages—but US medical privacy regulations do not give him that option.

Finally, what about aggregation and anonymization? Even in the medical case, most individuals would probably accept having their individual cases factored into some larger and appropriately blinded counts of diagnoses and treatments, for the greater social good. However, even this noble sentiment hides a minefield: what is "appropriate blinding"? Cynical cryptographic colleagues often lament that reports of anonymization are greatly exaggerated.

Policy tools

Even in the legacy IoC, the problem of specifying an *access policy*—whether Alice should be allowed to see *X* under conditions *Y*—is vexing. Way back in 1976, Harrison, Ruzzo, and Ullman proved that the problem "do these policy rules allow anything bad to ever happen?" (for a fairly simple and formal definition of "bad") was computationally undecidable in the general case. Way back in the early days of the World Wide Web, scientist Lorie Cranor crafted the Privacy Bird tool to help ordinary users ensure that websites respected their privacy wishes—only to find that while users had subtle and nuanced privacy preferences, developing a language in which they could specify them was extremely complicated. Later work by Cranor should how apparently minor choices in file access policy language can make it easy or hard for ordinary users to correctly express their desired behavior. See [52] for some of my ranting in this *access control hygiene* space.

When media (such as text and songs) started merging with the IoC world, the challenge of *digital rights management*—how to express appropriate usage rights in a way that machines can enforce—became another nightmare problem, still echoing today (e.g., see *https://www.eff.org/issues/drm*). The human world had the concept of "fair use," but it's hard to tell computers what that means.

When we go from the IoC to the IoT, we add even more dimensions to the privacy policy problem. Do users fully understand all the players that are involved in an interconnected cloud-supported service? Do users understand the implications of large aggregations of personal data? (For example, in the Dartmouth StudentLife project [39], test subjects who were willing to share some smartphone data would probably not have been willing to share their GPAs and mental health diagnoses—even though the former correlated with the latter.) What about things that last longer than the companies that support them or the people to whom they belonged? Can my wife still see her health record even though her name changed? Can my children listen to my music after I die?

As we've seen earlier in this chapter, IoT and IoC applications may often share more data with more parties than intended. Perhaps it would be useful to develop tools to "fuzz-test" policy: for instance, if *P* turns to *P′* due to some standard blunders, is *P′* still acceptable?

LAW AND STANDARDS

Industry and legal standards can be another vector to address "greater social good" concerns such as privacy. We already see this happening in the IoT space.

Even back in 2013, researchers in medical informatics lamented the lack of formal privacy policies in mHealth applications [42]. In February 2015, the office of Senator Edward J. Markey released a report looking at privacy in the IT-enhanced car [35] and described several disturbing findings, including:

> *Nearly 100% of cars on the market include wireless technologies that could pose vulnerabilities to hacking or privacy intrusions.*
>
> *Automobile manufacturers collect large amounts of data on driving history and vehicle performance....*
>
> *most do not describe effective means to secure the data....*
>
> *Customers are often not explicitly made aware of data collection and, when they are, they often cannot opt out without disabling valuable features, such as navigation.*

The report urged the US National Highway Traffic Safety Administration (NHTSA) and Federal Trade Commission to take action.

Around the same time, Edith Ramirez, chair of the FTC, warned [33]:

> *Connected devices that provide increased convenience and improve health services are also collecting, transmitting, storing, and often sharing vast amounts of consumer data, some of it highly personal, thereby creating a number of privacy risks.*

She urges the following:

> *(1) adopting "security by design"; (2) engaging in data minimization; and (3) increasing transparency and providing consumers with notice and choice for unexpected data uses*

In August 2015, the Online Trust Alliance (OTA) proposed a set of IoT rules to promote privacy [49]. Many of these will sound welcome to the reader who's read this far and seen the consequences of bad authentication, bad encryption, flawed interfaces, unpatchability, and lifetime troubles:

1. *Default passwords must be prompted to be reset or changed on first use or uniquely generated....*
2. *All user sites must adhere to SSL*[1] *best practices using industry standard testing mechanisms....*
3. *All device sites and cloud services must utilize HTTPS encryption by default.*
4. *Manufacturers must conduct penetration testing for devices, applications and services....*
5. *Manufacturers must have capabilities to remediate vulnerabilities in a prompt and reliable manner.*
6. *All updates, patches, revisions, etc. must be signed/verified.*

1 As one colleague notes, we hope they mean TLS, as (strictly speaking) SSL is obsolete.

7. *Manufacturers must provide a mechanism for the transfer of ownership including providing updates for consumer notices and access to documentation and support.*

Although it does not seem to have any ability to force compliance, the OTA is led by the heavy hitters of consumer IT (e.g., Microsoft) and web security infrastructure (e.g., DigiCert and Verisign), so one hopes it has clout. The proposed IoT Trust Framework has since gone through three more revisions and is now complemented by a "Consumer IoT Security and Privacy Checklist."

Also in summer 2015, the Healthcare Information Technology Policy Committee (HITPC), under the umbrella of the US federal efforts for electronic health, put forth a proposed set of rules for its space so that "patients should not be surprised about or harmed by collections, uses or disclosures of their information" [27].

TECHNOLOGICAL ENFORCEMENT

Even if we specify what "correct" behavior is (and perhaps have laws and standards to drive and guide that specification), how do we (again, as individuals or as a society) trust that our smart things actually follow these rules?

Consider two recent examples. In the Vizio analysis mentioned earlier, researchers discovered shenanigans [38]:

> *From this, it is obvious that the same data is being sent to Cognitive Networks servers through UDP and HTTP. This data is the fingerprint of what you're watching being sent through the Internet to Cognitive Networks. This data is sent regardless of whether you agree to the privacy policy and terms of service when first configuring the TV.*

Ars Technica made similar observations about Windows 10 [4]:

> *Windows 10 uses the Internet a lot to support many of its features. The operating system also sports numerous knobs to twiddle that are supposed to disable most of these features and the potentially privacy-compromising connections that go with them.*
>
> *Unfortunately for privacy advocates, these controls don't appear to be sufficient to completely prevent the operating system from going online and communicating with Microsoft's servers.*

> *For example, even with Cortana and searching the Web from the Start menu disabled, opening Start and typing will send a request to www.bing.com to request a file called threshold.appcache which appears to contain some Cortana information, even though Cortana is disabled. The request for this file appears to contain a random machine ID that persists across reboots.*

When I was a young computer scientist in industry taking a product through federal security validation, I was surprised to have to demonstrate not just that my product did what I said it would do, but also that it did not do what it did not say it did. I objected, but in hindsight, I see the wisdom.

Achieving this is hard, though. The long history of *side-channel* analysis shows that computing devices can communicate in ways observers (and owners) may not expect; applied cryptography shows how these communications can be made practically indistinguishable from random noise; and Scott Craver's *Underhanded C Contest* shows how even direct inspection of source code can fail to reveal what it really does. (Maybe the issues described in "Cryptographic Decay" on page 116 are in fact positive features: if you wait enough decades, you can finally decrypt your device's sneaky espionage reports.)

Works Cited

1. S. Ackerman and S. Thielman, "US intelligence chief: We might use the Internet of Things to spy on you," *The Guardian*, February 9, 2016.
2. American Academy of Periodontology, *Healthy Gums and a Healthy Heart: The Perio-Cardio Connection*. June 1, 2009.
3. BinaryEdge, "Data, technologies and security—Part 1," *blog.binaryedge.io*, August 10, 2015.
4. P. Bright, "Even when told not to, Windows 10 just can't stop talking to Microsoft," *Ars Technica*, August 12, 2015.
5. J. M. Broder, "Stalled out on Tesla's electric highway," *The New York Times*, February 8, 2013.
6. D. Burke, "Are tighter rules needed on recording devices in cars?," *CBC News*, May 24, 2016.

7. K. Carlon, "Runkeeper is secretly tracking you around the clock and sending your data to advertisers," *Android Authority*, May 13, 2016.

8. T. Chivers, "The Get Out Clause, Manchester stars of CCTV," *The Telegraph*, May 8, 2008.

9. L. Constantin, "Samsung smart TVs don't encrypt the voice data they collect," *ITworld*, February 18, 2015.

10. L. Constantin, "Smart TVs raise privacy concerns," *ITworld*, February 9, 2015.

11. J. Cook, "Apple is preparing to launch a voicemail service that will use Siri to transcribe your messages," *Business Insider*, August 3, 2015.

12. A. Coyne, "Westfield ditches SMS feature over privacy issues," *iTnews*, February 3, 2016.

13. M. Donoghue, "Arraignment delayed in fatal car–bike crash," *Burlington Free Press*, June 25, 2015.

14. S. Dunlap, "Drones could be used in Macon-Bibb for emergency response," *The Telegraph*, July 13, 2015.

15. J. Foster, "Someone gained access to private PLQ meetings, very easily," *CJAD News*, June 17, 2016.

16. L. Franceschi-Bicchierai, "One of the largest hacks yet exposes data on hundreds of thousands of kids," *Motherboard*, November 27, 2015.

17. S. Gallagher, "OwnStar: Researcher hijacks remote access to OnStar," *Ars Technica*, July 30, 2015.

18. S. Gallagher, 'EPIC' fail—How OPM hackers tapped the mother lode of espionage data," *Ars Technica*, June 21, 2015.

19. S. Gallagher, "OnStar gives Volt owners what they want: Their data, in the cloud," *Ars Technica*, November 25, 2012.

20. J. M. Gitlin, "Another driver says Tesla's Autopilot failed to brake; Tesla says otherwise," *Ars Technica*, May 13, 2016.

21. R. Giwargis, "San Jose looks at using garbage haulers to catch car thieves," *The Mercury News*, August 19, 2015.

22. D. Goodin, "Man-in-the-middle attack on Vizio TVs coughs up owners' viewing habits," *Ars Technica*, November 11, 2015.

23. R. Grenoble, "Tesla, New York Times still feuding over Model S review: Elon Musk releases data, reviewer counters," *The Huffington Post*, February 14, 2013.

24. B. Hambright, "Woman staged 'rape' scene with knife, vodka, called 9-1-1, police say," *LancasterOnline*, June 19, 2015.

25. D. He and others, "Security concerns in Android mHealth apps," in *Proceedings of the American Medical Informatics Association Annual Symposium*, November 2014.

26. K. Hill, "If you use Waze, hackers can stalk you," *Fusion*, April 26, 2016.

27. Health IT Policy Committee Privacy and Security Workgroup, *Health Big Data Recommendations*, August 11, 2015.

28. S. Knope, "October 4, 1984 and Libby Zion: The day medicine changed forever," *The Pearl*, November 7, 2013.

29. D. Kravets, "Worker fired for disabling GPS app that tracked her 24 hours a day," *Ars Technica*, May 11, 2015.

30. B. Krebs, "This is why people fear the 'Internet of Things,'" *Krebs on Security*, February 18, 2016.

31. R. Lawler, "Vizio IPO plan shows how its TVs track what you're watching," *Engadget*, July 24, 2015.

32. A. Liptak, "Major ruling shields privacy of cellphones," *The New York Times*, June 25, 2014.

33. N. Lomas, "The FTC warns Internet of Things businesses to bake in privacy and security," *TechCrunch*, January 8, 2015.

34. S. Margaritelli, "Nike+ FuelBand SE BLE protocol reversed," *evilsocket.net*, January 29, 2015.

35. Staff of E. Markey, *Tracking & Hacking: Security & Privacy Gaps Put American Drivers at Risk*. Office of the United States Senator for Massachusetts, February 2015.

36. C. Marlow, *Building a Mass Surveillance Infrastructure Out of Light Bulbs*. American Civil Liberties Union, July 23, 2015.

37. J. McLaughlin, "NSA looking to exploit Internet of Things, including biomedical devices, official says," *The Intercept*, June 10, 2016.

38. A. McSorley, "The anatomy of an IoT hack," *Avast Blog*, November 11, 2015.

39. M. Mirhashem, "Stressed out? Your smartphone could know even before you do," *New Republic*, September 22, 2014.

40. T. Mogg, "Hit-and-run suspect arrested after her own car calls cops," *Digital Trends*, December 7, 2015.

41. C. Moraff, "DEA wants inside your medical records to fight the war on drugs," *The Daily Beast*, June 10, 2016.

42. J. Mottl, "Mobile app privacy practices scarce, lack transparency," *FierceHealthcare*, August 24, 2014.

43. D. Munro, "Data breaches in healthcare totaled over 112 million records in 2015," *Forbes*, December 31, 2015.

44. E. Musk, "A most peculiar test drive," *Tesla Blog*, February 13, 2013.

45. M. Novak, "The FBI can neither confirm nor deny wiretapping your Amazon Echo," *Paleofuture*, May 11, 2016.

46. *NU.nl*, "Bedrijven mogen gezondheid medewerkers niet volgen via wearables," March 8, 2016.

47. Office of Public Affairs, Bureau of Consumer Protection, *Provider of Medical Transcript Services Settles FTC Charges That It Failed to Adequately Protect Consumers' Personal Information*. Federal Trade Commission, January 31, 2014.

48. P. H. O'Neill, "New research suggests that hackers can track subway riders through their phones," *The Daily Dot*, May 25, 2015.

49. Online Trust Alliance, *IoT Trust Framework—Discussion Draft*, August 11, 2015.

50. F. Y. Rashid, "Home invasion? 3 fears about Google Home," *InfoWorld*, June 15, 2016.

51. B. Schrier, "The Internet of First Responder Things (IoFRT)," *The Chief Seattle Geek Blog*, May 25, 2015.

52. S. Sinclair and S. W. Smith, "What's wrong with access control in the real world?," *IEEE Security and Privacy*, July/August 2010.

53. J. F. Smith, "Cyberattack exposes I.R.S. tax returns," *The New York Times*, May 26, 2015.

54. S. W. Smith, "Security and cognitive bias: Exploring the role of the mind," *IEEE Security and Privacy*, September/October 2012.

55. I. Thomson, "Hello Barbie: Hang on, this Wi-Fi doll records your child's voice?," *The Register*, February 19, 2015.

56. S. Trudeau and others, "The effects of introspection on creating privacy policy," in *Proceedings of the 8th ACM Workshop on Privacy in the Electronic Society*, November 2009.

57. Y. Wang and others, "Access control hygiene and the empathy gap in medical IT," in *Proceedings of the 3rd USENIX Conference on Health Security and Privacy*, 2012.

58. The White House, *Big Data: Seizing Opportunities, Preserving Values*. Executive Office of the President, May 2014.

59. C. Wong, "Sports wearables may affect athletes' privacy, paycheques as well as performance," *IT Business*, July 13, 2015.

7 Business, Things, and Risks

The business case necessary for IoT deployment may not necessarily align with IoT safety.

Deploying the IoT requires scale: lots of devices, distributed broadly. Given the capitalist slant of international society today, the actors who will make such deployment happen will be business entities, acting only when they see some business advantage. This central role of business entities and motivations will shape how the IoT unfolds. This chapter considers some of the resulting risks to society. The profit motivation:

- Can have direct risks for end users
- Can have privacy risks
- Can lead to arguably worse technology choices

How the IoT Changes Business

When discussing computer science aspects of the IoT, one often needs to say what's different. How does this new thing change the game? The same question should be asked when it comes to discussing the business aspects of the IoT.

DISRUPTING BUSINESS OPERATIONS

Over the last two years, the *Harvard Business Review* published a nice set of articles examining how the IoT changes traditional business operations. Looking at IoT and competition in 2014, Porter and Heppelmann stressed the advantage of the IoT *technology stack*: product linking to connectivity linking to the *product cloud.* Backend analytics on this aggregate data provides competitive advantages. This stack changes what businesses do [19]:

> *This opens the door to new competitors, such as the "productless" OnFarm, which is successfully competing with traditional agricultural equipment makers to provide services to farmers through collecting data on multiple types of farm equipment to help growers make better decisions, avoiding the need to be an equipment manufacturer at all....*
>
> *The basis of competition thus shifts from the functionality of a discrete product to the performance of the broader product system, in which the firm is just one actor*

GE CEO Jeff Immelt is quoted as observing that "every industrial company will become a software company."

In a companion piece on "digital ubiquity," Iansiti and Lakhani emphasize the business potential of remote sensors and backend analytics in the *industrial internet* [12]:

> *GE was at increasing risk of losing many of its top customers to nontraditional competitors...[who] aimed to shift the customer value proposition away from acquiring reliable industrial equipment to deriving new efficiencies and other benefits through advanced analytics and algorithms based on the data generated by that equipment.*

GE responded by using the industrial internet to change its business model:

> *Now revenue from its jet engines, for example, is tied not to a simple sales transaction but to performance improvements: less downtime and more miles flown over the course of a year.*

Iansiti and Lakhani also note the scalability of digital objects: "exact replication infinite times at zero marginal cost."

In a follow-up piece in 2015, Porter and Heppelmann further emphasized the business advantages of the industrial internet [20]:

> *To better understand the rich data generated by smart, connected products, companies are also beginning to deploy a tool called a* digital twin.... *Originally conceived by the Defense Advanced Research Projects Agency (DARPA), a digital twin is a 3-D virtual-reality replica of a physical product.*

> *As data streams in, the twin evolves to reflect how the physical product has been altered and used and the environmental conditions to which it has been exposed.*

Porter and Heppelmann also note how the easy malleability of software (compared to hardware) changes the design and development process from slow, discrete cycles to something more continuous:

> *In conventional products, variability is costly because it requires variation in physical parts. But the software in smart, connected products makes variability far cheaper. For example, John Deere used to manufacture multiple versions of engines, each providing a different level of horsepower. It now can alter the horsepower of a standard physical engine using software alone.*

Porter and Heppelmann describe this as *evergreen design*. (Of course, this approach gives rise to a risk: aftermarket hacking that lets customers get more power without paying the manufacturer for it.)

Similar points from a different perspective can be found in IoT promotional information from vendors such as IBM and Intel. In fact, Intel has a case study on the use of the IoT to market craft beer—something that made a nice A/V supplement for an undergraduate class (see Figure 7-1).

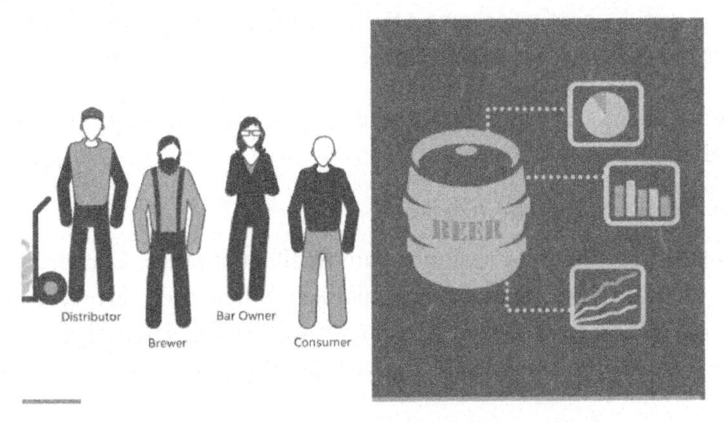

Figure 7-1. Intel's SteadyServ iKeg case study demonstrates the transformative power of IoT on business, in a domain readily appreciated by college students. (Image reproduced with the permission of Intel Corporation, which owns the copyright.)

DISRUPTING THE PROFIT PARADIGM

"Paradigm disruption" is a term that's sure to become overused when applied to the IoT. Nonetheless, quantum changes in how things get done can change how money flows for those services—and, as a consequence, change the nature of industries.

As a case in point, one need only look at what's happened to the recorded music industry in the US in the last 30 years. In 1986, music was sold primarily packaged in the vinyl LP format. The newer CD format was beginning to catch on; older formats such as the cassette tape and the eight-track tape still existed, but were fading. Music reached the ears of prospective purchasers via radio stations broadcast over actual radio waves; stations logged what music they played, and this log data fed back to royalty payments to musicians (or at least songwriters).

The emergence of the IoC changed all this. Vinyl and tape vanished (although vinyl is making a small comeback in hipster circles); CDs mostly vanished too. (The "used record" stores I would regularly visit when visiting college towns became "used CD" stores, and then disappeared altogether.) To a great extent, radio waves have been replaced by internet streaming—and often not from a traditional radio station but from a program running on a Pandora server. Lamenting these changes, Jonathan Taplin (in the *New York Times*) offered the fascinating observation [24]:

> *In 2015, vinyl record sales generated more income for music creators than the billions of music streams on YouTube and its competitors.*

In this one area alone, technology disrupted. What further industry disruption will the IoT bring?

"Napster moments" in car insurance

One area that's received discussion is automobile insurance. Currently, drivers of cars pay premiums to insurance companies, which then handle most of the expenses of car accidents. Presumably by using careful actuarial modeling to set rates, the insurance companies have mastered the statistical odds and stayed in business.

However, consider the IoT vision where the on-road fleet is replaced (slowly or quickly) by smart vehicles that drive themselves. Most technologists accept that sooner or later, self-driving vehicles will indeed be safer than traditional ones, at least in the aggregate of fewer accidents and damages overall (although not necessarily in the fate of any one specific driver). If this vision comes to fruition, will we still need to pay large insurance premiums—and if we pay only small ones, what happens to the insurance companies?

As *Bloomberg* put it in 2015 [2]:

> *The auto insurance industry is having its Napster moment. Like record companies at the dawn of online music file sharing, Allstate, Geico, State Farm, and others are grappling with innovations that could put a huge dent in their revenue.*

They quote Warren Buffett, indirect owner of Geico:

> *"If you could come up with anything involved in driving that cut accidents by 30 percent, 40 percent, 50 percent, that would be wonderful," he said at a conference in March. "But we would not be holding a party at our insurance company."*

Some companies are exploiting the safety edge of smart vehicles. SiliconValley.com reports [1]:

> *Liberty Mutual...is offering discounted rates on cars with assistive features such as blind-spot warnings and "collision preparation" systems that tighten seat belts and perform other safety enhancements if the vehicle's systems detect an imminent crash.*

Others are shifting focus areas. The *Christian Science Monitor* noted [3]:

> *One possibility could be for the insurance giant [State Farm] to reinvent itself as a "life management company," as the company put it in a patent application recently published by the U.S. Patent Office.... [T]he company could analyze data about a customer's vehicles, home and personal health, find patterns and offer "personalized recommendations, insurance discounts, and other added values or services that the individual can use to better manage and improve his or her life."*

"GOOGLE MOMENTS"?

On the other hand, as the purpose of insurance is to reduce the maximum costs any one individual faces from a bad but statistically unlikely event by spreading the costs across a larger population, the IoT may in fact create opportunity.

Recall from Chapter 1 how the IoT triggers two general rants from security analysts:

- The IoT increases the attack surface (making attacks more likely).
- The IoT amplifies the physical consequences of an attack.

Taken together, these tenets suggest a potential for the IoT to create new bad events requiring insurance protection. Smart, connected vehicles may be exposed to malicious remote manipulation; we may have fewer crashes from driver error but more from electronic vandalism. Similarly, smart, connected homes may make it easier for a burglar to determine which houses have valuable appliances and when these houses are likely empty, or for a vandal to maliciously manipulate heating and cooling and ovens—hence, the IoT might bring more burglaries and vandal-induced fires.

For these opportunities to pay off, however, insurers need to get a handle on the statistics to ensure the events are indeed sufficiently unlikely across a pool of customers. Security researchers have long mourned the dearth of effective *security metrics*—how do we measure the actual risk from the unknown zero-days in the infrastructure? Insurance—even for cyberattacks in the IoC, let alone in the emerging IoT—may thus help with this problem. Another potential benefit might be the emergence of standards (as our society sees now with things such as fire safety codes governing construction) to reduce the aggregate risk, put in place by entities whose profit motivation ensures these standards work.

This may already be happening. In May 2015 *The Security Ledger* reported [18] that Columbia Casualty Insurance had claimed in court that it could deny coverage for a privacy spill at Cottage Health Systems because the insured:

> *Failed to follow "minimum required practices," as spelled out in the policy. Among other things, Cottage "stored medical records on a system that was fully accessible to the internet but failed to install encryption or take other security measures to protect patient information from becoming available to anyone who 'surfed' the Internet."*

Profit and Safety

How much is one life worth? To slightly paraphrase the Bible, what would it profit a person to gain all the world's riches but lose himself or herself in the process?

Framing these questions this way implies that, obviously, rational actors would always choose safety over wealth. On a consumer level, no cost is too high to pay; on a producer level, no profit justifies causing damage.

Unfortunately, this is not how humans operate.

IN HISTORY

When security researchers (such as I) discuss the risks of the coming IoT, listeners often hear it as endless litany of things that can go wrong and say, "Since we can't solve these problems, should we just give up on building this smart future?" The security researcher then usually gives a more hopeful response, along the lines of how identifying problem areas can focus attention on solutions or mitigation strategies so that we can have the smart future without the smart risks.

But there's another possibility as well.

Suppose a group of researchers proposed a new technology, *X*, that promised to fundamentally change and improve (mostly) society in all sort of exciting ways. However, *X* has some downsides:

- In the US alone, its adoption would directly kill over 30,000 people each year—and indirectly hurt far more.
- In the US alone, its adoption would near-permanently consume landmass whose area exceeds the state of Georgia.
- Worldwide, its adoption would fundamentally alter the climate of the Earth (in a bad way) and create destabilizing geopolitics, leading to wars, despotism, and terrorism.

One would imagine that, given these downsides, society would either reject adoption of *X* as simply not worth the cost, or perhaps pursue a more prudent path of delaying deployment until we had used our collective ingenuity to eliminate these hazards. The idea that society would happily accept *X* with these costs —to the point where life without *X* would be unimaginable—is unthinkable.

And yet, this is what society has chosen with the automobile: we accepted both the transformative new technology and all of its costs. Perhaps this will be the fate of the IoT as well.

We (as a society) know how to build safer cars, but we (as a society) choose not to. With cars—or with any other potentially dangerous products, such as lead paint or pesticides or lawnmowers, or with difficult social choices such as the high cost of implementing positive train control in the US even though it would have eliminated deadly derailments—individual consumers choose not to pay the higher cost for statistical safety; manufacturers choose a different point on the profit-aggregate safety curve. (Of course, more subtleties may lurk here; for example, the existence of positive control might make some engineers even more likely to engage in risky behavior.)

IN THE IOT

Will society make the same "irrational" choices in the IoT?

The IoT has already seen choices between profit and cost. In 2014, the *Harvard Business Review* noted [19]:

> *In residential water heaters, A.O. Smith has developed capabilities for fault monitoring and notification, but water heaters are so long-lived and reliable that few households are willing to pay enough for these features to justify their current cost.*

In May 2015, *Fusion* discussed a recent video posted from Central America [11]:

> *A group of people stand in a garage watching and filming a grey Volvo XC60 that backs up, stops, and then accelerates toward the group. It smashes into two people, and causes the person filming the video with his phone to drop it and run. It is terrifying.*

Apparently, what was going on is that some smart Volvos have a feature called *pedestrian detection*. The people in the video were apparently trying to demonstrate that. However:

> *It appears that the people who bought this Volvo did not pay for the "Pedestrian detection functionality," which is a feature that* costs more money.

The opening of this chapter discussed the amorphous nature of smart products: instead of having *N* different variations of a product, a vendor can produce one generic one that becomes one of *N* depending on which software is installed —or perhaps even on which software, already present, is enabled. In the case of the scary Volvo, it's possible that the car in question actually knew how to avoid the pedestrians, but a flag was turned off because the owner hadn't paid. Someone—the vendor, or the owner, or both—decided that the added benefit wasn't worth the additional price or lost revenue, even though the marginal cost may very well have been zero. (Indeed, a decade ago, a scientist who does digital forensics for law enforcement told me of an alleged drug dealer who had used strong encryption on his computer but neglected to pay the shareware license, so it was still running in crippled, low-security mode when seized.)

Other differentiating aspects of the IoT may also lead to similar choices. The ability for an IoT device to grant remote access to physical reality (thermostats? heart devices?) introduces choices between convenience and risk. The ability of highly instrumented reality to generate massive amounts of data could enable backend analytics to detect all sorts of potentially useful things, including high-accuracy predictions of bad things about to happen. Pointing this microscope at some targets means not pointing it at others. As with the possibility of identifying the 9/11 actors before the attack, it will be awkward if we have to look back and say we had the pieces to predict the bridge collapse or the imminent heart attack but did not, because vendors or consumers, probably motivated by money somehow, made different choices about services.

IN THE HUMAN MIND

As Chapter 6 discussed, psychology has identified (and can experimentally demonstrate) all sorts of cognitive bias scenarios where human minds confidently make choices that would appear to be irrational based on data. Humans believe driving is safer than flying, insist on expensive but not highly useful healthcare for some situations (e.g., patients at end of life) but in other scenarios decline cheaper choices with higher utility (e.g., vaccination), or worry about attacks in public restrooms from transsexuals but not from congressmen. Psychologists have also looked at ways (e.g., [14, 25]) to reframe these questions so human minds make decisions that are "better," in the sense that they do not look so irrational in hindsight or from third-party judgment.

The business of the IoT opens up new vistas for choices and mixes lots of apples and oranges. It might be worth exploring the implications for human decision making:

- On an individual level (what would a human choose?)
- On a business level (what trade-offs will be chosen among revenues and costs?)
- On a social level (when and how should nations and international coalitions step in and regulate to shape what the market forces do?)

When the User Is the Product

As the opening of this chapter discussed, players from IBM to Intel to the *Harvard Business Review* stress the potential of big data analytics as a business driver for the IoT. Vastly increasing the types and numbers of measurements can translate to increased knowledge, which can translate to a competitive advantage.

However, this scenario can lead to an interesting inversion, where the end user of technology is not the consumer. From a financial point of view, the consumers are the large entities paying for aggregate data about the users. As a popular saying puts it, "When the product is free, then you are the product."

IN HISTORY

To see this pattern in the current IoC, one need only look at antitrust arguments against Google. For example, Nathan Newman in the *Huffington Post* writes [16]:

> *The pleasant experience of using Google products is little different (in any economic analysis) from the pleasant massage administered to Kobe beef cattle in Japan; each is just a tool to increase the quality of the product delivered up to the real customers...Here's the key place to start in understanding proper technology policy for Google: there is no market for search engines; there is no market for online geolocation mapping software; there is no market for online video. Google, by making these products free, has destroyed those markets in favor of an alternative economic model of selling individual attention and precise information about those users to advertisers. You are the product, not the customer.*

Massive data can enable more accurate identification: potential customers for advertisers and potential employees for employers. This effectiveness can be creepy—witness the father who discovered his daughter was pregnant via Target coupons mailed to her [5], or the services of the British startup Score Assured [4]:

The company wants to, in the words of co-founder Steve Thornhill, "take a deep dive into private social media profiles" and sell what it finds there to everyone from prospective dates to employers and landlords.

As a scientist and also a former low-level competitive cyclist, I've read about the science and technology behind doping, and this idea of using big data analytics to focus narrowly on individuals reminds me of doping: it's creepy, but it likely works.

IN THE IOT

Businesses are already seeing the advantage of harvesting the personal data the IoT will collect. Earlier in this chapter, we saw how State Farm was contemplating switching from selling insurance for traditional cars to selling analytics from data collected by smart cars. For another example, consider Google's 2014 purchase of Nest for $3.2 billion. Analysts such as *ITworld*'s Matthew Mombrea observed that Google must have had something else in mind: "to recover its purchase price, Google would have to sell a lot" of the $250 smart thermostats [15]. Mombrea hypothesized the secondary market of selling efficiency services to the power grid. Others saw the purchase as an effort by Google to use the IoT to gather yet more data about users in order to enhance its targeted advertising business [10]:

But years ago it moved its ambitions into being a universal data collector not only to power its advertising business through intimate customer knowledge, but also to serve as a common fabric for all services to use. After all, the more services you use, the more Google can discern about you.

However, more recently, pundits have been puzzling about how and why Nest/Google's sparkly future has fizzled.

For yet another example of the business of using the IoT to harvest individual data, consider Fitbit fitness wearables. At first glance, one would think the consumers for these devices and services would be the individuals who choose to buy and wear them. However, *ITworld* reports how Fitbit is also marketing to corporations [17]:

"We think virtually every company will incorporate fitness trackers into their corporate wellness programs," Fitbit CFO Bill Zerella said Tuesday.

> *Businesses are using wellness programs to increase employee productivity, decrease the number of sick days workers take and potentially reduce health care costs, Zerella said during a session at the Pacific Crest Global Technology Leadership Forum.*

Fitbit's corporate thrust includes not just selling Fitbits to employees, but also selling special software services (available via special portals) to their employers. (Although as Chapter 6 observed, this sort of thing is considered a privacy violation in the Netherlands.)

Corporations are also using the data avalanche to monitor and tune employee performance more directly. In 2015, the *New York Times* surveyed various products in this space, including tools at GE to "give workers instant feedback from bosses and colleagues," Amazon's Anytime Feedback, and Sapience's Buddy [23]:

> *Khiv Singh, a Sapience vice president, noted that data surrounded workers: "We have pedometers to measure how far we walk, apps to monitor our blood pressure, stress level, the calories we're taking in, the calories we're burning. But the office is where we spend the majority of time, and we don't measure our work."*

One of Sapience's customers "was surprised by what he found" when his company started using Buddy:

> *"Engineers would write on their time sheets that they were doing development for eight hours, but we started to see a very different set of activities that people are performing," Mr. Bohra said. "Meetings. Personal time. Uncategorized time. Performing research on something that maybe already should be a part of our knowledge repository."*

To be fair, the *Times* also notes that BetterWorks tries to make employees happier rather than employers.

What does this mean for society? Personally, I find this observation from a Score Assured founder troubling rather than assuring:

> *"If you're living a normal life...then, frankly, you have nothing to worry about."*

Profit and Technological Choices

Selling and buying play a central role in shaping which technologies permeate society. The principal forces behind selling and buying (in theory, at least) are the enlightened self-interest of manufacturers looking to make a profit and of consumers looking for good value. Do these forces lead to good, effective technology?

IN HISTORY

History gives many scenarios where the answer to this question was "no."

Greengard quotes internet pioneer Vint Cerf on the state of IT a century ago [9]:

> *We certainly didn't want to wind up with a situation parallel to the 1910s and 1920s, when a business had a dozen different telephones sitting on a desk—all using a different proprietary system and requiring a person to know which telephone service to use to reach someone else.*

From Vint's point of view, society would be better off with interoperable IT—except the profit motive alone hasn't given us that yet. Decades later, the VHS videotape format won out over Betamax, which some argue had been better technology. In the early days of the web, Netscape's SSL won out over "Secure HTTP," which some argued had been the better choice. (We could also go decades earlier and examine the Edison/Tesla feud between DC and AC electric power.)

The recent Oracle/Google litigation shows another example of the potential impact of profit motives and technology. From Oracle's point of view, Google violated copyright law by using Java APIs that were the property of Oracle; from Google's point of view (the one supported by the latest court decision), this was fair use. One could argue that Oracle initiating this legal action was a sensible business move: the APIs were its property, and Oracle deserved compensation for use of that property. However, many in the industry feared that an Oracle victory would fundamentally change IT, in a negative way. For example, Klint Finley in *Wired* wrote [6]:

> *Nothing less is [at] stake than the future of programming.... Regardless of how the jury rules, the case has already had a permanent effect on the way developers build software.... [S]ince the appeals court has already ruled that APIs are subject to copyright, that could open a whole new fron-*

tier of lawsuits aimed at startups and open source projects that have copied APIs in order to ensure their products are compatible with popular commercial products.

SCO's earlier litigation against various commercial users of Linux might be a similar example, except widespread opinion in the technical community was that it had no merit, so it's not as interesting. (However, in industry in the 1990s, I personally saw how such intellectual property squabbles contributed to a reluctance to use open source software—which in turn led to redundant work and new bugs.)

A Reddit user posted an account of a more short sighted case of profit versus technology: how a power plant employed workarounds to avoid paying software license fees, almost leading to a shutdown [13].

Another angle to consider here is the relative timing of new technology and new business ventures framing that technology. Why did Facebook succeed but Myspace disappear? Everyone knows about Google, but who remembers Lycos and AltaVista?

IN THE IOT

The preceding examples where local business self-interest potentially missteered the global state of IT all began in the IoC. It's hard to cite an IoT-specific example so far, except perhaps for Philips blocking third-party lightbulbs from its smart lighting hubs (see Chapter 8). However, it's also hard to believe that the same trends will not continue. The IoT offers the same multiplicities of choices and possibly proprietary APIs and services. IoT-time will move even faster than IoC-time—of which Milton Friedman warned [7]:

Is it really in the self-interest of Silicon Valley to set the government on Microsoft? Your industry, the computer industry, moves so much more rapidly than the legal process, that by the time this suit is over, who knows what the shape of the industry will be.

Furthermore, for massive deployment to happen, a large class of IoT things will need to be much smaller and cheaper than IoC things [22]:

The FTC also cautions that many devices are inexpensive or "disposable," essentially calling into question whether the threat assessment and internal productivity outweighs any reward of consistently patching new attack vectors each time one is discovered.

HACKING AND BUSINESS

September 2016 brought news [21] of a perhaps novel combination of IoT security, hacking, and business:

> *When hackers at cybersecurity startup MedSec Holdings discovered security vulnerabilities in St. Jude Medical pacemakers and defibrillators, they contacted Carson Block, who runs investment firm Muddy Waters Capital. MedSec and Block struck an unprecedented partnership: The hackers provided data showing the devices, used by tens of thousands of people, had life-threatening flaws; and Block bet against St. Jude Medical stock by selling it short, agreeing to pay MedSec fees based on how much St. Jude's shares fell. If the shares didn't fall, MedSec would be out the money for its research and other upfront costs.*

Hackers found holes, and used the stock market both to punish those responsible for the holes and to finance their work.

Businesses and Things and People

The preceding sections all focused on ways in which the IoT changes businesses, and the way in which business motivations may negatively impact the rollout of the IoT.

The relationships and lifetimes of the businesses responsible for IoT technology may also interact pessimally with the behavior and lifetime of the technology itself. For one example, the publicity surrounding the radio-based attacks in Daimler Chrysler's Jeep Cherokee overlooked the question of which other automobile manufacturers may have sourced radio technology from the same vendor. For another, Chapter 1 noted how the "penetrate and patch" model will not defend us when the patchers are long out of business. *The Register* extends these concerns to other aspects of a business's technology stack [8]:

> *Don't expect your paid service from a big provider or a start up will still be there in five years' time.*

Clearly, the IoT will need the energy stemming from the enlightened self-interest of profit-oriented businesses in order to happen. Mitigating some of the risks that can result may require harnessing the various tools—public policy, public and private standards, consortia, consumer action—such enlightened self-interest has tuned and shaped in the past.

Works Cited

1. E. Baron, "Self-driving cars to disrupt auto insurance industry," *SiliconValley.com*, June 19, 2016.
2. N. Buhayar and P. Robison, "Can the insurance industry survive driverless cars?," *Bloomberg Businessweek*, July 30, 2015.
3. A. Danise, "Will driverless cars mean the end of auto insurance?," *The Christian Science Monitor*, January 23, 2016.
4. C. Dewey, "Creepy startup will help landlords, employers and online dates strip-mine intimate data from your Facebook page," *The Washington Post*, June 9, 2016.
5. C. Duhigg, "How companies learn your secrets," *The New York Times Magazine*, February 16, 2012.
6. K. Finley, "The Oracle–Google case will decide the future of software," *Wired*, May 23, 2016.
7. M. Friedman, "The business community's suicidal impulse," *Cato Policy Report*, March/April 1999.
8. S. Gilbertson, "If you're not paying, you're product: If you ARE paying, it's no better," *The Register*, November 4, 2013.
9. S. Greengard, *The Internet of Things*. MIT Press, 2015.
10. G. Gruman, "Google's grand plan for Nest goes way beyond the Internet of Things," *InfoWorld*, January 14, 2014.
11. K. Hill, "Volvo says horrible 'self-parking car accident' happened because driver didn't have 'pedestrian detection,'" *Fusion*, May 26, 2015.
12. M. Iansiti and K. R. Lakhani, "Digital ubiquity: How connections, sensors, and data are revolutionizing business," *Harvard Business Review*, November 2014.
13. ITDepartmentOfOne, "Never trust a subcontractor," *Reddit*, July 21, 2015.
14. M. Liersch and C. McKenzie, "Duration neglect by numbers—And its elimination by graphs," *Organizational Behavior and Human Decision Processes*, 2009.

15. M. Mombrea, "Google's real plan behind the purchase of the Nest thermostat," *ITworld*, April 25, 2014.

16. N. Newman, "You're not Google's customer—You're the product: Antitrust in a Web 2.0 world," *The Huffington Post*, May 29, 2011.

17. F. O'Connor, "Fitbit caters to corporations, and not just with discounted fitness trackers," *ITworld*, August 11, 2015.

18. Paul, "Clueless clause: Insurer cites lax security in challenge to Cottage Health claim," *The Security Ledger*, May 26, 2015.

19. M. E. Porter and J. E. Heppelmann, "How smart, connected products are transforming competition," *Harvard Business Review*, November 2014.

20. M. E. Porter and J. E. Heppelmann, "How smart, connected products are transforming companies," *Harvard Business Review*, October 2015.

21. J. Roberson and M. Riley, "How hackers used pacemaker vulnerabilities to play the market," *Bloomberg Businessweek*, September 5, 2016.

22. C. Rouland, "FTC report on IoT: The debate over opportunity, liability, and privacy," *Bastile Blog*, February 10, 2015.

23. D. Streitfeld, "Data-crunching is coming to help your boss manage your time," *The New York Times*, August 17, 2015.

24. J. Taplin, "Do you love music? Silicon Valley doesn't," *The New York Times*, May 20, 2016.

25. P. E. Tetlock, "Thinking the unthinkable: Sacred values and taboo cognitions," *TRENDS in Cognitive Sciences*, July 2003.

Laws, Society, and Things

The IoT crosses many boundaries (cultural, jurisdictional, and national); laws and management of the IoT will also need to cross these boundaries.

Law[1] is undeniably another large force in shaping how things happen in this world. The IoT traverses not only geographical lines separating governmental jurisdictions, but also domains of life previously covered by separate customs. This boundary crossing leads to interesting interactions between the IoT and law, and this chapter surveys some of the principal ones:

- The use of smart technology to hide from the law
- The use of law to keep smart technology from being scrutinized
- How the IoT introduces new things, neither fish nor fowl, that create challenges for what legal framework they should inherit

When Technology Evades Law

In theory, law governs behavior by sanctioning behaviors deemed to be sufficiently bad for the social contract. However, in the IoC and IoT, we have already seen scenarios where the behavior of smart technology has somehow sidestepped this governance. Let's consider a few.

1 And regulation and public policy and such.

CASE STUDY: VW EMISSIONS

In the Garden of Eden story from Genesis, eating from the Tree of Knowledge enables human to sin. A cynic might predict that in the IoT story, making things smart will enable them to sin: to use the adaptiveness and resourcefulness of computing to do objectively bad things, such as deceptively cheat. Unfortunately, this has already happened.

Powering internal combustion engines with diesel instead of gasoline offers many efficiency advantages. However, burning diesel can generate more pollution. In recent years, diesel-powered cars have become popular in Europe; they also made inroads in the US, as cars equipped with "clean diesel" technology passed the strict pollution standards of the Environmental Protection Agency (EPA).

Some researchers were curious. Eric Niler in *Wired* writes [21]:

> *In 2013, a small non-profit group decided to compare diesel emissions from European cars, which are notoriously high, with the US versions of the same vehicles. A team led by Drew Kodjak, executive director of the International Council on Clean Transportation, worked with emissions researchers at West Virginia University to test three four-cylinder 2.0-liter diesel cars in the Los Angeles area: a Jetta, a Passat, and a BMW.*

The EPA tests evaluate emissions when the car is stationary. The ICCT/WVU team instrumented the cars to evaluate emissions when actually driving on roads. Figure 8-1 summarizes the result: the VW diesel cars ("Vehicle A" and "Vehicle B") somehow emitted far more pollutants when being driven on roads than when being tested while stationary. The stationary tests were under the EPA threshold, but the actual road tests far exceeded it.

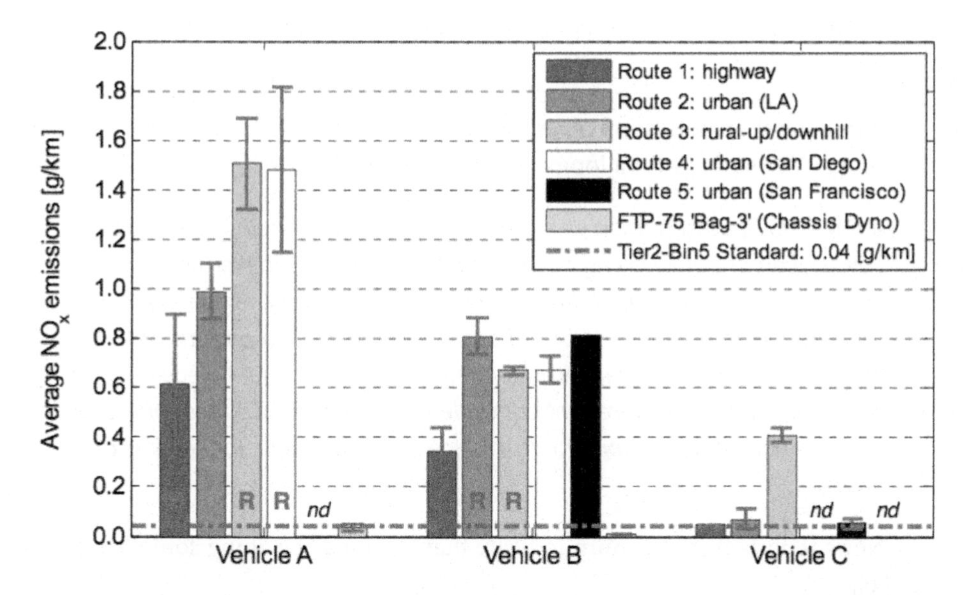

Figure 8-1. This graph (from the ICCT/WVU study [29]) shows the pollution generated by three diesel cars when being driven on roads (the "Route" bars) against the pollution generated during stationary EPA tests (the "FTP-75" bars). The left two vehicles are VWs. (Used with permission.)

This difference was puzzling, and a long conversation began [26]:

VW engineers continued to suggest technical reasons for the test results. None of the explanations satisfied regulators, who indicated the models wouldn't be certified.

However, then the bombshell dropped:

"Only then did VW admit it had designed and installed a defeat device in these vehicles in the form of a sophisticated software algorithm that detected when a vehicle was undergoing emissions testing," the EPA said in its letter to VW.

Wired provides more details about this "sophisticated software algorithm":

Computer sensors monitored the steering column. Under normal driving conditions, the column oscillates as the driver negotiates turns. But during emissions testing, the wheels of the car move, but the steering wheel doesn't. That seems to have have been the signal for the "defeat device"

> *to turn the catalytic scrubber up to full power, allowing the car to pass the test.*

Subsequent investigation suggests that this smart circumvention was an intentional VW strategy for a long time. The *New York Times* reports [12]:

> *A PowerPoint presentation was prepared by a top technology executive at Volkswagen in 2006, laying out in detail how the automaker could cheat on emissions tests in the United States.... In a laboratory, regulators would try to replicate a variety of conditions on the road. The pattern of those tests, the presentation said, was entirely predictable. And a piece of code embedded in the software that controlled the engine could recognize that pattern, activating equipment to reduce emissions just for testing purposes.*

News reports from Belgium indicate that Opel may be doing something similar [25].

For Volkswagen, consequences included a $14.7 billion settlement in the US, with actions still pending in other countries. Researchers also concluded that the cheating software led to deaths [34]:

> *According to the model, the extra NOx from VW's cars [in the US] will cause about 10 to 150 (or a median of 59) people to die 10 to 20 years early. Hospital bills and other social costs add up to $450 million.*

For the Internet of Cheating Things, this may be only the tip of the iceberg. Considering other potential examples is left as an exercise for the reader—although "Things 'on the witness stand'" on page 180 gives a few.

A good way to ensure that things are not cheating is to ensure that their internals can be scrutinized, as "When Law Stops Scrutiny of Technology" on page 171 will discuss.

"WEASEL WORDS"

In the US and elsewhere, the legal framework codifies a notion of negligence. If something bad happens to Alice because of a basic flaw in something Bob designed and built, then Bob should be held responsible.

It's long been observed that somehow the software industry has avoided this doctrine—vendors of software systems sidestep responsibility for negligent behavior in a way that would make vendors of automobiles, airbags, and hot coffee jealous. It's tempting to posit a general negligence principle: hard things require care, but soft things are blame-free.

In the IoT, the soft things are deeply permeating the space of the hard things. Unfortunately, we are also starting to see the negligence principle permeate as well.

Privacy spills

For one example, consider the privacy spill from toy company VTech's backend servers, discussed back in Chapter 6. *Motherboard* quoted the president of VTech assuring customers, "We are committed to the privacy and protection of the information you entrust with VTech" [15]. However, *Motherboard* also noted that VTech's terms and conditions now include this advisory (in all caps):

> *YOU ACKNOWLEDGE AND AGREE THAT ANY INFORMATION YOU SEND OR RECEIVE DURING YOUR USE OF THE SITE MAY NOT BE SECURE AND MAY BE INTERCEPTED OR LATER ACQUIRED BY UNAUTHORIZED PARTIES.*

When it comes to legal issues, one can usually find opposing points of view. *Motherboard* further notes:

> *Rik Ferguson, the vice president of security research at Trend Micro, said the clause is "outrageous, unforgivable, ignorant, opportunistic, and indefensible," and likened it to "weasel words." Despite this surprising change —a British law professors told me he's "never seen a clause like that before"—legal experts doubt the provision has any real value.*

Which view will dominate?

Smart medicine

In the domain of smart health, one can find examples with more dire consequences. My colleague Harold Thimbleby cites the Mersey Burns app (approved by the UK NHS), which helps a clinician determine how much fluid a burn victim needs based on the extent of their burns. Harold notes first that the app's legal warranty removes responsibility from the vendor and regulator:

You agree to indemnify and hold...the NHS harmless from any claim...as a result of your use or misuse of the app.

Harold notes further that the warranty allows itself to change arbitrarily:

The NHS may modify this disclaimer...at any time...without giving notice to you.

In conversation, Harold asks what the reader is likely asking now: what good is a written warranty whose terms can change at any point without notice?

Unfortunately, the Mersey Burns case is not simply a case of weasel words. The application prompts the clinician to input the extent of a patient's burns first on the front side of the body, and then on the back side of the body. The app also gives the clinician two ways to enter these measurements: graphically, or via a percentage. Harold identified a dangerous bug: indicating severe burns on the front side graphically followed by indicating minor burns on the back via percentage somehow causes the app to forget about the fluids required because of the front injuries.

Patients might die from this bug, but somehow no one is responsible.

My colleague Ross Koppel has long lamented how this pattern is endemic in health IT (at least in the US), and specifically calls out clauses commonly found in the contracts hospitals have with the vendors of health IT systems [19]:

One clause prohibited clinicians from publicly displaying screen shots... even if they felt those screen shots illustrate a danger to patient safety.... A part of that clause also prohibited clinicians from speaking pejoratively about the vendor's product. The second clause, "hold harmless," said that the vendor was not responsible for any errors committed because of their products even if the vendor had been repeatedly informed that the product was defective in some way.... The legal logic of the clause is that the vendor merely creates a "tool" used by a learned intermediary...[who] has the authority to take whatever information is shown and make a considered professional judgment, including realizing that the information shown to him or her via the software is incorrect.

The legal disclaimers say the vendor is not responsible, the vendor does not have to fix bugs, and the clinicians are not allowed to disclose those bugs. Sadly,

Ross documents many cases were patients were harmed—and even killed—because of such bugs.

In the years since Ross first published these concerns, other organizations have joined the call to eliminate these clauses.

When Law Stops Scrutiny of Technology

The scientific world preaches the value of *peer review* and *reproducible results*: advances are not considered valid unless they can be carefully examined and verified. Engineering preaches the value of *closed loop* systems: not just acting, but measuring the result and adaptively tuning. Even the basic 20th-century American mythos celebrates tinkering: the computer giant that started in a garage, the young woman wrenching on an old car.

A common thread through all of this is the ability for individuals to examine technology. However, even in the IoC, we began to see situations where law was used to discourage such examination. (Noted security and public policy researcher Ed Felten even titled his blog *Freedom to Tinker* in response to this situation.)

For example, May 2016 brought news of a particularly ironic case: in Florida, security researcher David Levin—working *with* local elections supervisor Dan Sinclair—demonstrated a security hole in a state elections website. The government response was to arrest him [24]:

> *"Dave didn't cause these problems, he only reported them," Sinclair said, adding that the elections office could not previously detect intrusions. Levin also provided defensive measures to the state about how it could fix the hole and detect further intrusions.*

June brought news that the ACLU had "filed a lawsuit with the US Department of Justice contending that the Computer Fraud and Abuse Act's (CFAA's) criminal prohibitions have created a barrier for those wishing to conduct research and anti-discrimination testing online" [5], because the act is used to "criminalize violations of websites' 'terms of service'" [3]:

> *The CFAA violates the First Amendment because it limits everyone, including academics and journalists, from gathering the publicly available information necessary to understand and speak about online discrimination.*

Closer to the IoT space, in 2012 researchers Roel Verdult, Flavio Garcia, and Baris Ege discovered flaws in the wireless key/starter protocol used by many vehicles, including Volkswagens. Their result was to be published at the 2013 USENIX Security Symposium, but VW used the UK courts to stop publication [2]:

> *VW and Thales argued that the algorithm was confidential information, and whoever had released it on the net had probably done so illegally. Furthermore, they said, there was good reason to believe that criminal gangs would try to take advantage of the revelation to steal vehicles.*

Two years later, the paper was finally published [17]. It is interesting to note the involvement of VW (which at the same time was using IT to evade EPA rules) in stopping scrutiny of its IT.

Unfortunately, this wasn't the first time that a USENIX Security research paper was delayed due to legal action. In 2000, the Secure Digital Music Initiative—an industry consortium focused on digital rights management (DRM)—held a challenge for researchers to scrutinize various DRM techniques. Researchers from Princeton (including Ed Felten, later to start the *Freedom to Tinker* blog) and Rice were largely successful [7], but court action by SDMI delayed publication of their results.

CASE STUDY: THE DMCA

The Digital Millennium Copyright Act (DMCA) is a 1998 US law often invoked by players trying to suppress scrutiny and decried by advocates for the freedom to tinker. As the Electronic Frontier Foundation (*https://www.eff.org/issues/dmca-rulemaking*) puts it:

> *The Digital Millennium Copyright Act prohibits "circumventing" digital rights management (DRM) and other "technological measures" used to protect copyrighted works. While this ban was meant to deter copyright infringement, many have misused the law to chill competition, free speech, and fair use.*

In the IT space, the ostensible intention of the DMCA was to protect as intellectual property the software embedded in devices. If a pirate—or researcher or curious tinkerer—"circumvented" a manufacturer's barrier (no matter how small) to examine this software, that could be interpreted as a violation of the DMCA.

What behaviors the DMCA actually prohibits has been a matter of ongoing contention. Nonetheless, it's hard to dispute the chilling effect. In the VW emissions scandal discussed earlier, discovery that VWs were programmed to provide acceptable levels of pollution only when they detected they were being tested happened indirectly, because researchers tried an alternate way of measuring pollution. If researchers could have looked directly at the code, that discovery might have happened much earlier.[1] The EFF argues [31]:

> *Automakers argue that it's unlawful for independent researchers to look at the code that controls vehicles without the manufacturer's permission.... The legal uncertainly created by the Digital Millennium Copyright Act also makes it easier for manufacturers to conceal intentional wrongdoing.... Volkswagen had already programmed an entire fleet of vehicles to conceal how much pollution they generated, resulting in a real, quantifiable impact on the environment and human health. This code was shielded from watchdogs' investigation by the anti-circumvention provision of the DMCA.*

(Personally, as someone who spoke out against the DMCA when it was first enacted, I found it ironic to hear it being discussed on NPR nearly two decades later.)

One example of the evolving contention over what the DMCA prohibits is *jailbreaking* cellphones. The Copyright Office (part of the Library of Congress) determined that jailbreaking a cellphone violated the DMCA and was prohibited; it took federal legislation in 2014 to make it legal again [6].

In *Slate*, Kyle Weins then observed in January 2015 [33]:

> *How many people does it take to fix a tractor?.... [I]t actually takes an army of copyright lawyers, dozens of representatives from U.S. government agencies, an official hearing, hundreds of pages of legal briefs, and nearly a year of waiting. Waiting for the Copyright Office to make a decision about whether people like me can repair, modify, or hack their own stuff.*

1 However, as the Underhanded C Contest shows, it's possible to hide malicious behavior in code that looks very innocent.

As the Copyright Office considered, manufacturers offered an opposing view [32]:

> *John Deere—the world's largest agricultural machinery maker—told the Copyright Office that farmers don't own their tractors. Because computer code snakes through the DNA of modern tractors, farmers receive "an implied license for the life of the vehicle to operate the vehicle."*
>
> *It's John Deere's tractor, folks. You're just driving it.*

In October 2015, the Copyright Office (FEDREG) ruled that "computer programs that are contained in and control the functioning of a motorized land" were exempt from the DMCA "when circumvention is a necessary step undertaken by the authorized owner of the vehicle to allow the diagnosis, repair or lawful modification of a vehicle function" and also "for the purpose of good-faith security research."

Weins, however, worries about the IoT's slippery slope [33]:

> *Phones are just the beginning. Thanks to the "smart" revolution, our appliances, watches, fridges, and televisions have gotten a computer-aided intelligence boost. But where there are computers, there is also copyrighted software, and where there is copyrighted software, there are often software locks. Under Section 1201 of the DMCA, you can't pick that lock without permission. Even if you have no intention of pirating the software. Even if you just want to modify the programming or repair something you own.*

Churn continues. In December 2015, *Techdirt* reported that Philips had released a firmware update that would remove the ability of purchasers of the Philips Hue smart lighting bridge from using third-party lightbulbs [9]. Presumably, circumventing these restrictions would be considered a copyright violation.

In May 2016, state legislators in Michigan (historical home of the US automobile industry) bypassed the DMCA altogether and introduced legislation that would make hacking into a car a crime, possibly meriting life imprisonment. In *ComputerWorld*, Darlene Storm responded [28]:

Of course we don't want to wait until hackers are remotely taking control and crashing cars before we figure out what should be done to malicious attackers, but if security researchers can't look for vulnerabilities without fear of life in prison, then aren't we all less safe?

Would we have been better off not to know that hackers could remotely seize control of a Jeep as it is speeding down the highway? I don't think so.

I concur.

When New Things Don't Fit Old Paradigms

A common motif in this book is that layering networked IT on top of previously dumb objects creates new sorts of things whose existence and behaviors raise new questions—including legal ones, when these new beasts, neither fish nor fowl, do not fit into the standard paradigms. Even way back at the dawn of the IoC, sharp-eyed legal scholars pointed out that putting a click-through license on the front of an "electronic" book changed the governance of reader/publisher behavior from copyright law to contract law, which was somehow worse for the reader. The new age of the IoT brings new dilemmas.

SEND IN THE DRONES

For an extreme example of adding networked computing (and a degree of autonomy) to a previously dumb thing, a teenager in 2015 attached a gun to a drone and posted a video of it on YouTube. CNN reported [20]:

The gun drone in Connecticut appears to have been fired on private property and—so far, authorities said—it did not appear any laws were broken.... "It appears to be a case of technology surpassing current legislation," police in Clinton, Connecticut, said.

In 2016, *Popular Science* reported on a Finnish project that equipped a drone with a chainsaw [1]. This effort also yielded an amusing video—but, sadly, did not give rise to a dry observation from police on whether it was illegal. Bringing things back to cyber, in 2015 the *Intercept* reported that an Italian hacking company was in discussions with a Boeing drone subsidiary to explore using drones to deliver malware by hovering over targets and intercepting wireless communications [8].

LICENSE TO SELF-DRIVE

Another standard item in the IoT vision is replacing our population of cars with a fleet of self-driving vehicles that can drive more safely and efficiently. One of the upsides this vision presents is freeing humans from having to commit time and attention to driving: instead, commuting time becomes work or relaxation time; alcohol-impaired humans can still "drive" home safely; humans with disabilities (e.g., vision impairment) preventing them from driving traditional cars can become autonomously mobile.

However, thanks to at least a century of evolution, driving is surrounded by sociocultural processes with foundations in law. When the car becomes the driver, what should happen with these processes and laws?

For one example, consider licensing. In the US and many other nations, humans need to obtain (and then carry) a government-issued license before driving a vehicle on public roads. Should the human who is not driving a self-driving car also require a license?

- At first glance, this question sounds ridiculous. Of course not—the human is not driving!
- On the other hand, what if the car provides the ability for a human driver to take control, perhaps as a safety feature? In this case, perhaps we do need licenses after all—but then what happens to the upsides of freeing humans from driving themselves?
- Even without the ability for a human to take over fully, a self-driving car may still take some control from its human—for instance, for destination, route options, speed preferences, etc. In this case, should we still require some kind of licensing (or at least some minimum age)? If there are multiple passengers, which ones require licenses?

In summer 2015, the UK Department of Transport released *The Pathway to Driverless Cars: A Code of Practice for Testing* outlining legal guidelines for this transition [10]. This document distinguished between a *test driver* ("the person who is seated in the vehicle in a position where they are able to control the speed and direction using manual controls at any time") and a *test operator* ("someone who oversees testing of an automated vehicle without necessarily being seated in the vehicle"). However:

> *The test driver or test operator must hold the appropriate category of driving licence for the vehicle under test, if testing on a public road. This is true even if testing a vehicle's ability to operate entirely in an automated mode. It is strongly recommended that the licence holder also has several years' experience of driving the relevant category of vehicle...*
>
> *Test drivers and operators should remain alert and ready to intervene if necessary throughout the test period.*

Another legal/social aspect of driving is insurance. Accidents will happen. How do we handle who pays for the damage? Indeed, in most places in the US (but not New Hampshire, where I live), the biggest obstacle for young drivers is neither earning the license nor buying an inexpensive used car—rather, it is obtaining the legally required insurance. With traditional vehicles, we (as a society) have worked out a system for financial responsibility for car accidents that works, mostly—there are still cases of lawsuits about negligent manufacturers of cars and servers of alcohol, and (in New Hampshire specifically) the problem of uninsured drivers. How does this system translate to when vehicles somewhat or mostly or entirely drive themselves (Figure 8-2)?

Figure 8-2. In this XKCD strip, Randall Monroe points out the difficulty of identifying the driver of a self-driving car. (Source: xkcd.com.)

Reporting on discussions between Google and the UK on testing self-driving cars, the *Telegraph* also observed [30]:

> *Google has taken a special interest in the thorny issue of how driverless cars will be insured. Because a computer program, rather than a human,*

> *would be controlling the vehicle, experts have suggested that manufacturers will be held responsible.*

In *IEEE Spectrum*, Nathan Greenblatt went even further [18]:

> *It is the year 2023, and for the first time, a self-driving car navigating city streets strikes and kills a pedestrian. A lawsuit is sure to follow. But exactly what laws will apply? Nobody knows.*

(Chapter 7 discussed a related problem: if autonomous vehicles have fewer accidents, what happens to the business case for selling insurance?)

Yet another standard aspect of our legal/social driving practice is interaction with law enforcement. A driver being pulled over by police has been a standard cultural motif since the days of the *Keystone Cops* movies. Law also dictates specific driver behaviors when encountering other privileged vehicles, such as fire trucks and ambulances and school buses. How should these behaviors translate to self-driving cars?

Will Oremus of *Slate* repeated a scenario from RAND [23]:

> *The police officer directing traffic in the intersection could see the car barreling toward him and the occupant looking down at his smartphone. Officer Rodriguez gestured for the car to stop, and the self-driving vehicle rolled to a halt behind the crosswalk.*

As a thought experiment, this may sound reasonable. But Oremus considers the slippery slope:

> *If a police officer can command a self-driving car to pull over for his own safety and that of others on the road, can he do the same if he suspects the passenger of a crime? And what if the passenger doesn't want the car to stop—can she override the command, or does the police officer have ultimate control?*

These scenarios make me think of technical challenges reminiscent of Chapter 1 and Chapter 5. How do we (as a society) set up an authentication infrastructure that permits all the vehicles from different manufacturers and countries to verify "Stop, police!" commands from officials of all the different types of law enforcement agencies? Given the historical tendency of interfaces to unintentionally permit too much power, what will happen if someone other than authorized law enforcement personnel can do this a vehicle? Given also the unfortunate historical tendency of some law enforcement officers to overreach their authority, what will happen when a rogue officer does this?

HEALTHY ENTERTAINMENT

Smartphones have provided a ubiquitous networked platform for IoT-style applications. Many of these applications are medical, ranging from directly measuring properties of the user's body (e.g., heart rate) to helping the user track and manage other medical-related issues, such as calories consumed or anxiety/depression incidents.

In 2014, Brian Dolan noticed something interesting [11]:

> *This week MobiHealthNews sought out apps in Apple's AppStore that appeared to pitch themselves as useful medical or health-focused apps, but also included some iteration of that common legal disclaimer: "For entertainment purposes only". While none of the apps we found appear to be trying to take advantage of their users with a fantastical claim...the inclusion of the "entertainment" disclaimer is still a bit puzzling....*
>
> *After all, how entertaining is a medical calculator app that helps you figure out the stages of a patient's acute tubular necrosis? I'm not a doctor. I've never attempted such a calculation myself. But I'm guessing it's not particularly fun.*

The working assumption here is that installing a medical application on a smartphone makes it a *medical device* and thus brings it under the purview of the US FDA, whose mission includes regulating and certifying medical devices in order to ensure the safety of patients and effectiveness of treatments. What's troubling here is the emergence of this class of apps, which are clearly being used for medical purposes but sidestepping oversight. The FDA is trying to address this problem [13].

A related issue here is how to dovetail the need to continually update software (to fix the inevitable bugs, security and otherwise—recall Chapter 1) with the need for FDA certification on medical devices. A commonly held view is that vendors and clinicians are loath to patch software on medical devices because doing so would require taking the devices through FDA certification again. Medical security researcher Kevin Fu disputes that recertification is always required but discusses the difficult gray area [16]:

> *Guidance documents are peppered with conditional language open to interpretation. If you want to scare a regulatory affairs specialist, just add a bunch of implicit "if" or "unless" conditional branch statements.... [T]he absolute claim that "FDA rules prevent software security patches" is false. But there is a half truth hiding behind the sentiment. The rules are sufficiently fuzzy to cause misunderstandings and unintended interpretations.*

Chapter 1 quoted from a Bloomberg article on the penetration tests performed by Billy Rios and others on medical devices at the Mayo Clinic [27]. The same article reported on subsequent testing Rios performed on the Hospira infusion pump. Rios found holes: "He could set the machine to dump an entire vial of medication into a patient." However, Rios could not get regulators to pay attention: "The FDA seems to literally be waiting for someone to be killed before they can say, 'OK, yeah, this is something we need to worry about.'" These concerns came full circle when Rios himself was hospitalized—and connected to a Hospira infusion pump.

Eventually, the FDA did issue a warning with specific guidance for how healthcare facilities could reduce the risk from these holes. Although perhaps less than the dire, "Fix this now!" requirement a security analyst might have wanted, this action did demonstrate a welcomed transition in regulatory culture for medical devices—from "Patching for security is bad" to "Patching for security is good."

THINGS "ON THE WITNESS STAND"

Chapter 6 discussed the potential for IoT devices to betray their owners. Law enforcement officials are already using data from such devices as part of investigations, and it's not hard to foresee a day when such data is used in court; your devices may testify against you.

However, this sort of thing has been happening for a while (the future has been here before!). Data from speed traps, red light cameras, and breathalyzers has been used in legal proceedings for decades. In recent years, the US has seen a spate of copyright infringement lawsuits launched by the recording industry based on computer-generated data.

Looking at the copyright cases in particular, hackers Sergey Bratus and Anna Shubina, working with law professor Ashlyn Lembree, explored the issues that arise when things start testifying on the witness stand [4]:

> *Thus it appears that the* only *entity to "witness" the alleged violations and to produce an account of them for the court—in the form of a series of print-outs—was in fact an autonomous piece of software, programmed by a company acting on behalf of the plaintiffs and RIAA, and running on a computer controlled by this company.*

One can find incidents where bugs in breathalyzers and speed trap devices have resulted in cases being dismissed because the bugs rendered the output unreliable. Bratus, Lembree, and Shubina cite a particular case where a developer intentionally introduced the error to increase revenue from traffic tickets; the annual Underhanded C Contest shows how effectively a malicious programmer can hide evil behavior in code. So, rationally, one must conclude that testimony from things should not be automatically trusted. However, have legal mechanisms caught up to this reality?

> *Witnesses in court make their statements under oath, with severe consequences of deviating from the truth in their testimony. Witnesses are then cross-examined in order to expose any biases or conflicts of interest they might have. Computer-generated evidence comes from an entity that cannot take an oath...nor receive an adversarial examination....*
>
> *In short, a human witness' testimony is not automatically assumed to be trustworthy. Specific court procedures such as cross-examination and deposition by the opposing lawyers have evolved for challenging such testimony.*

The authors of this report point out that developing such procedures to scrutinize testimony from things is still a matter for "research by legal scholars." Fortunately, there have been several cases where US courts have required things like breathalyzers to undergo code reviews, and Bratus and Lembree themselves defended perhaps the only case where the Recording Industry Association of America (RIAA) withdrew its suit with prejudice—so some progress is happening.

Looking Forward

Moving from the IoC to the IoT puts us in interesting times.

This chapter discussed many ways in which legal issues have arguably increased the chance for a dangerous future with the IoT. However, we have also seen progress.

In January 2015, the US FTC issued a report outlining thinking on the IoT [14], discussing topics such as "how the long-standing Fair Information Practice Principles (FIPPs), which include such principles as notice, choice, access, accuracy, data minimization, security, and accountability, should apply to the IoT space'." The report recommended that ``companies should build security into their devices at the outset" and "continue to monitor products throughout the life cycle and, to the extent feasible, patch known vulnerabilities." The concept of *data minimization* was raised too: "Companies should limit the data they collect and retain, and dispose of it once they no longer need it."

The FTC report also wrestled with the balance of enacting new legislation versus working within existing regulatory frameworks, and with the balance of regulation versus innovation.

In the same month, Ofcom (the "communications regulator in the UK") issued its own report looking ahead to the IoT [22]. As might be expected from an office focused on communications, the report concentrated on issues such as networking and RF spectrum—but it did also address regulatory frameworks for privacy:

> *In so far as the IoT involves the collection and use of information identifying individuals, it will be regulated by existing legislation such as the Data Protection Act 1998. We have concluded that a common framework that allows consumers easily and transparently to authorise the conditions under which data collected by their devices is used and shared by others will be critical to future development of the IoT sector.*

Ofcom also stressed the need for *consumer literacy*:

> *Some respondents identified the benefit of advocating and communicating the potential benefits associated with the IoT more broadly. In particular, respondents noted the need to raise consumer awareness on how new devices and apps will be collecting and using personal data to deliver IoT services.*

As noted above, these are interesting times; one should expect many more developments in the legal and regulatory arenas.

Works Cited

1. K. Atherton, "Finnish filmmakers gave a drone a chainsaw," *Popular Science*, April 1, 2016.
2. BBC, "Car key immobiliser hack revelations blocked by UK court," *BBC News*, July 29, 2013.
3. E. Bhandari and R. Goodman, *ACLU Challenges Computer Crimes Law that Is Thwarting Research on Discrimination Online*. American Civil Liberties Union, June 29, 2016.
4. S. Bratus, A. Lembree, and A. Shubina, "Software on the witness stand: What should it take for us to trust it?," in *Proceedings of the Third International Conference on Trust and Trustworthy Computing*, 2010.
5. N. Cappella, "ACLU lawsuit challenges Computer Fraud and Abuse Act," *The Stack*, June 29, 2016.
6. R. Cox, "Senate passes bill to allow 'unlocking' cell phones," *The Hill*, July 15, 2014.
7. S. Craver and others, "Reading between the lines: Lessons from the SDMI challenge," in *Proceedings of the 10th USENIX Security Symposium*, 2001.
8. C. Currier, "Hacking team and Boeing subsidiary envisioned drones deploying spyware," *The Intercept*, July 18, 2015.
9. T. Cushing, "Light bulb DRM: Philips locks purchasers out of third-party bulbs with firmware update," *Techdirt*, December 14, 2015.
10. Department for Transport, *The Pathway to Driverless Cars: A Code of Practice for Testing*. February 2015.
11. B. Dolan, "The rise of the seemingly serious but 'just for entertainment purposes' medical app," *MobiHealthNews*, August 7, 2014.
12. J. Ewing, "VW presentation in '06 showed how to foil emissions tests," *The New York Times*, April 26, 2016.
13. FDA, CDRH, and CBER, *Mobile Medical Applications Guidance for Industry and Food and Drug Administration Staff*. U.S. Department of Health and Human Services Food and Drug Administration, Center for Devices and

Radiological Health, and Center for Biologic Evaluation and Research, February 9, 2015.

14. Federal Trade Commission, *Internet of Things: Privacy & Security in a Connected World*. FTC Staff Report, January 2015.

15. L. Franceschi-Bicchierai, "Hacked toy company VTech's TOS now says it's not liable for hacks," *Motherboard*, February 9, 2016.

16. K. Fu, *False: FDA does not allow software security patches*. October 17, 2012.

17. S. Gallagher, "Researchers reveal electronic car lock hack after 2-year injunction by Volkswagen," *Ars Technica*, October 12, 2015.

18. N. Greenblatt, "Self-driving cars will be ready before our laws are," *IEEE Spectrum*, January 19, 2016.

19. R. Koppel, "Great promises of healthcare information technology deliver less," in *Healthcare Information Management Systems: Cases, Strategies, and Solutions*, A. C. Weaver and others, Eds. Springer International Publishing, 2016.

20. M. Martinez and others, "Handgun-firing drone appears legal in video, but FAA, police probe further," *CNN*, July 21, 2015.

21. E. Niler, "VW could fool the EPA, but it couldn't trick chemistry," *Wired*, September 22, 2015.

22. Ofcom, *Promoting Investment and Innovation in the Internet of Things*. January 27, 2015.

23. W. Oremus, "Should cops be allowed to take control of self-driving cars?," *Slate*, August 24, 2015.

24. D. Pauli, "Researcher arrested after reporting pwnage hole in elections site," *The Register*, May 9, 2016.

25. L. Pauwels, "Are Opel dealers modyfing the software of polluting Zafiras?," *FlandersNews*, January 18, 2016.

26. J. Plungis and D. Hull, "VW's emissions cheating found by curious clean-air group," *Bloomberg*, September 19, 2015.

27. M. Reel and J. Robertson, "It's way too easy to hack the hospital," *Bloomberg Businessweek*, November 2015.

28. D. Storm, "Hack a car in Michigan, go to prison for life if new bill becomes law," *Computerworld*, May 2, 2016.

29. G. J. Thompson and others, *In-Use Emissions Testing of Light-Duty Diesel Vehicles in the United States*. Center for Alternative Fuels, Engines & Emissions, West Virginia University, May 15, 2014.

30. J. Titcomb, "Google's meetings with UK Government over driverless cars revealed," *The Telegraph*, December 12, 2015.

31. K. Walsh, *Researchers Could Have Uncovered Volkswagen's Emissions Cheat If Not Hindered by the DMCA*. Electronic Frontier Foundation, September 21, 2015.

32. K. Wiens, "We can't let John Deere destroy the very idea of ownership," *Wired*, April 21, 2015.

33. K. Wiens, "Before I can fix this tractor, we have to fix copyright law," *Slate*, January 13, 2016.

34. S. Zhang, "New study links VW's emissions cheating to 60 early deaths," *Wired*, October 30, 2015.

The Digital Divide and the IoT

The IoC already exacerbates class differences. Can we keep the IoT from doing the same?

The dawn of the web (and the original Internet of Computers) introduced concerns about the *digital divide*: how this new technology might amplify class differences, particularly because education and affluence appeared to be necessary to get on board in the first place. By reaching into more aspects of life and even into basic physical infrastructure, the IoT may also increase digital division. This chapter will consider:

- How digital divides emerged in the IoC
- How digital divides may continue in the IoT
- How digital divides may emerge when IT is required to support basic rights
- How IoT applications may enforce preexisting socioeconomic divides
- How IoT applications may create divides even among connected classes

How Digital Divides Emerged in the IoC

I remember the dawn of the web. Through the distorted lens of hindsight, it's easy to wax lyrical about the sudden emergence of pervasive client browsers, the common HTTP protocol—and the electricity as we all realized that this ability for ordinary citizens and enterprises to start exchanging information and interacting would change everything. Fast-forward now a quarter century, and one could add the conclusion: "See, we were right—it *did* change everything."

To some extent, these stories are true. At the beginning, there was an electric excitement—and I even had the chance, as a government security analyst, to work with a number of citizen-facing agencies that were exploring how this new technology could increase efficiency and access to services. To paraphrase a standard client vision:

> *We are legally mandated to provide information service X to citizens but we are only given a limited budget, so maybe this internet web thing can help us reach more citizens without having to build more offices. (Can you tell us what the security and technology risks are?)*

And from the perspective of anyone reading these words, "everything" did change: the internet and web are how we now read the news, shop, stay connected with our friends, consume audio and video media, do banking, pay bills—and also take care of various government services.

However, these rosy stories sneak in some implicit assumptions. Words such as "everything," "everyone," and "pervasive" imply universality—but they are really universal *only to the experience of the people telling the stories.*

In the early 1990s, the community I was a part of and whose excitement I shared about this new thing consisted mainly of computer science researchers affiliated with universities and laboratories that happened to be close to the network backbone—and where funding and circumstance left the researchers with then-fancy computers on their desktops and time to play with them for not-necessarily-work-related purposes. This small community was only a tiny sliver of society at large.

THE DIGITAL DIVIDE

As the decades passed, citizenship in this new world grew to encompass a larger fraction of the population, but a fraction still.

As Susan Crawford wrote in the *New York Times* in 2011 [3]:

> *Telecommunications, which in theory should bind us together, has often divided us in practice. Until the late 20th century, the divide split those with phone access and those without it. Then it was the Web: in 1995 the Commerce Department published its first look at the* digital divide, *find-*

ing stark racial, economic and geographic gaps between those who could get online and those who could not.

Many demographics do not share in the connectivity of the geek elite. In the US alone, even in 2015, the data showed differences. According to the White House [13]:

- The highest-income quartile has almost twice the internet usage as the lowest quartile (see Figure 9-1).
- Households where the head has finished college have almost twice the internet usage as households where the head has not finished high school.
- Households headed by whites and Asians have more usage than households headed by other races.
- Usage drops off with age (see also Figure 9-1).[1]

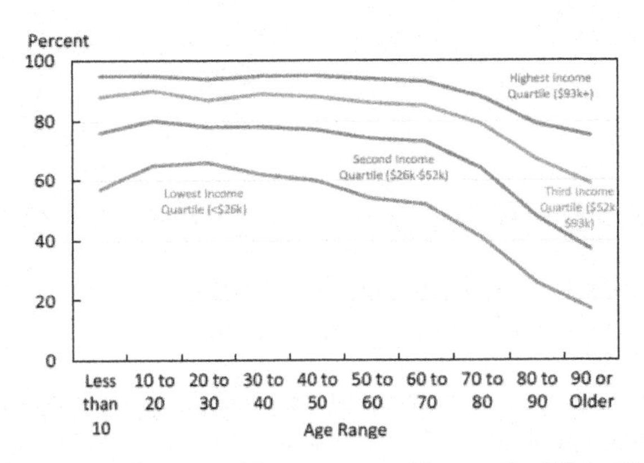

Figure 9-1. Home internet use by age and income, 2013. (Source: the White House [8].)

Again, even in the US alone, the data shows that more than just straightforward socioeconomic factors are at play. Internet usage strongly correlates with geography, although not always by rural/urban divisions (see Figure 9-2)—bandwidth may be a bigger factor [3]:

1 Of course, it can be challenging to find the causality behind this correlation. One older friend laments "millennials who are just not interested in becoming geeks."

While we still talk about "the" internet, we increasingly have two separate access marketplaces: high-speed wired and second-class wireless.

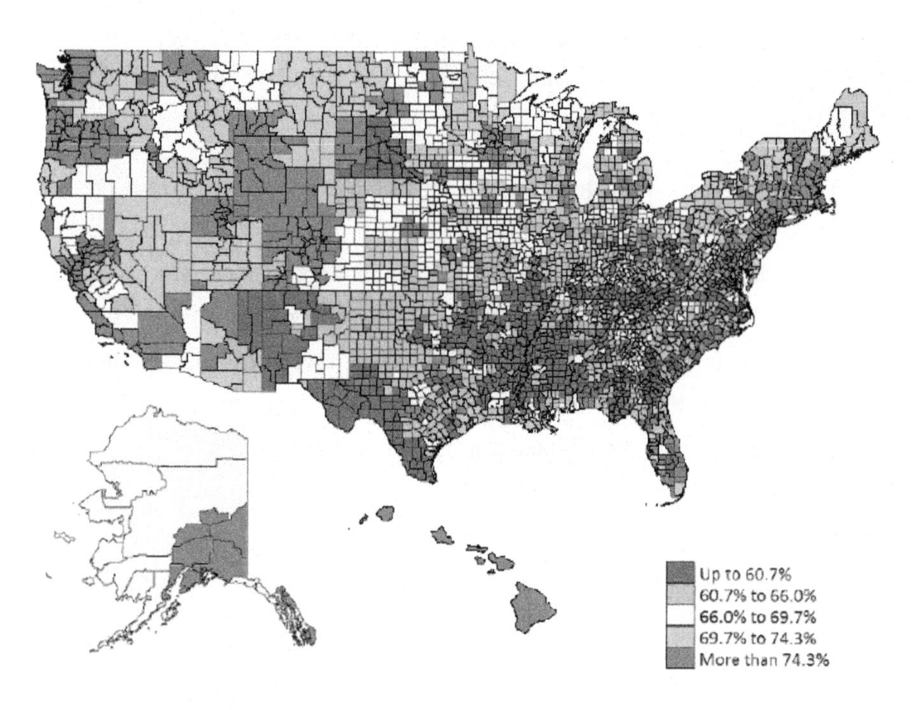

Figure 9-2. Internet adoption in the US in 2013. (Source: the White House [8].)

In 2013, *Mashable* shared this anecdote from a school in Newark, New Jersey ("a city with one of the highest poverty rates in the U.S.")[6]:

> *"If they knew someone who could turn their phone into a hot spot, they would actually pay other students to use their data," said Robert Fabriano, a 30-year-old teacher at the school. "They would trade bus tickets...if they lived two or more miles away, or a couple of bucks if they had it."*

School examples help reveal the slippery slope of the digital divide. What starts out as innocuous—for instance, the inability to share in the cultural experience of watching funny cat videos—quietly becomes a more substantial handicap: the inability to apply for jobs, the inability to do homework. The Pew Research Center quotes a university librarian [1]:

> *What I see are a handful of first-world white men touting their shiny new toys. Put this in context with someone struggling to get by on a daily basis —in the US or in other countries: what these devices primarily signify is a growing gulf between the tech haves and have-nots. That said, I'm not boycotting these devices—I see them as interesting and important. But just as students today are burdened if they don't have home Internet—and at the university where I work, that is true of some of our commuter students, much as people might find that hard to believe—there will be an expectation that successful living as a human will require being equipped with pricey accoutrements.... Reflecting on this makes me concerned that as the digital divide widens, people left behind will be increasingly invisible and increasingly seen as less than full humans.*

Even in the IoC, even when we consider a single developed country such as the US, we see a digital divide. If we extend internationally to countries with different stages of economic development, different types of government, and different levels of stability, the situation gets worse.

Other factors also play into the digital divide. Can a blind person be a full cyber-citizen? Can someone who does not speak the dominant language of the region they live in? In my own classroom, I was confronted with a challenge: how to enable a student who'd injured his dominant arm to still do his computer architecture project (where text-based workarounds such as dictation would not suffice).

How Digital Divides May Continue in the IoT

The IoC age itself has already created and exacerbated digital divides. Although they would likely not agree on the appropriate solution, most political ideologies would regard this as a problem—when segments of society are systematically cut off from the basic infrastructure of society. As IoC access becomes more important to education and economic advancement, the divide becomes even more troubling, since it becomes self-perpetuating.

As we rush to the IoT and distribute intelligence throughout the physical world, will we risk making the digital divide even bigger?

CONNECTIVITY TO MACHINES

One avenue to consider is the basic plumbing of network connectivity. The standard full vision of the IoT has the smart things talking back to the big data backend; where this channel does not exist or is constrained, the communication will be limited. Will the applications even work without this connectivity?

In a 2015 analysis on the digital divide, Huawei lamented [7]:

> *A lack of locally relevant, quality and accessible services for many users is limiting the benefits they can achieve through digital technologies. These are often the very people that could most benefit from these services: those who do not have quality education or healthcare systems, those with poor infrastructure and geographic difficulties, or those with poor eyesight, hearing or mobility. Though not always necessary, many services are built for—or operate best with—high internet speeds.*

Others have noted that the global (or at least multinational) aspect of many IoT applications will require a widespread IoT ecosystem to come to fruition—for example, Weber and Weber raise the example of RFID-enhanced smart shipping [12].

The security vision of the IoT stresses the need for pushing patches to the smart things (or bringing about a revolution in software engineering that eliminates that need). Lack of sufficient connectivity will reduce this ability. Will the shading in the map of Figure 9-2 also correspond to malware infections in the future IoT? (For that matter, what about in the present IoC?)

CONNECTIVITY BETWEEN PEOPLE

In May 2015, Mary Catherine O'Connor (in the *IoT Journal*, writing about the IoT and agriculture) identified a different digital divide, between the IoT technologists and the experts in the domains in which they're trying to embed the IoT [9]:

> *Technologists tend to be more excited about the IoT on the farm than farmers and chefs are.... [T]here is—at least in Central California—a divide between Silicon Valley and food producers.... [I]t seemed as though the subtext to questions and comments from farmers was: What do the folks in Silicon Valley know about how to produce food? Didn't all those VCs already do enough damage by over-hyping Internet technology before the dot-com bust in the early 2000s? What damage can they inflict on the ag industry?*

Indeed, an element of the history of computer science not often mentioned by computer scientists has been excessive optimism (or perhaps hubris) that real-world processes could be easily captured with just the proverbial small amount of programming (see Figure 9-3). One example of this excessive optimism was the prediction of early AI researchers that exact computational reproduction of the human thought process was imminent. Examples in the academy include complexity and computability theory: useful problems turn out to be uncomputable, or probably intractable (e.g., NP-complete), or apparently intractable with unknown foundation (e.g., factoring large integers). The potentially revolutionary (or potential dead end) field of quantum computing arose because of the apparent intractability of simulating quantum physics on classical computers. Less academic examples include the continued failure of digital rights management to capture the nuances of fair use (see Chapter 8), and the continued struggle of health IT and EMRs to capture the nuances of medical workflow (e.g., see [11]). Will the IoT bring information technology solutions to the world's problems—or just information technologists trying to solve them?

Figure 9-3. As this XKCD cartoon illustrates, not all real-world processes can be captured easily by software. (Source: xkcd.com.)

When IT Is Required to Support Basic Rights

In the context of humans, "rights" is a dangerous term to use. Discussion of rights threatens to bring in cans of worms: government structures, ideologies, social contracts, etc. For the purposes of this book, let's sidestep those issues and use the working definition that a "right" is something that society decides—formally or informally—a human merits simply by being a citizen, or perhaps even by just being human.

CERTIFICATES

Even in the world of bricks and mortar and paper, technology does not always nicely mesh with human and citizen rights. For example, in the US in recent decades, many social processes—such as boarding an airplane, or (more recently, in some states) voting—operate on the implicit assumption that everyone has a government-issued photo ID. However, as the Brennan Center for Justice at NYU reports, about 11 percent of Americans do not [4]:

> *Many seniors and many poor people don't drive. In big cities, many minorities rely on public transit. And many young adults, especially those in college, don't yet have licenses. A good number of these people, particularly seniors, function well with the IDs they have long had—such as Medicaid cards, Social Security cards or bank cards. Among the elderly, many of them have banked at the same branch for so long that tellers recognize them without needing to see their IDs.*

Hence the news has seen much discussion of laws requiring birth certificates and other documentation in order for a citizen to vote—and of the large numbers of people who have bona fide trouble obtaining this documentation. I personally have a relative who had trouble getting a copy of his birth certificate (due to lax bureaucracy in a New York mill town 80 years ago), a colleague who doesn't have one at all (due to migration from Cuba), and a spouse who has continual trouble flying because the name on her passport does not match the name on her driver's license.

So the simple idea of "use this special piece of paper" to support rights processes on a national scale has problems—11 percent of the US population has problems. If we can't manage paper certificates and IDs, then how are we going to manage digital credentials such as public key certificates? How are we going to manage citizen identification in the IoT without excluding segments of society?

(Indeed, how to authenticate a citizen was a concern for our egovernment explorers back in the 1990s. The IRS "Get Transcript" debacle described in Chapter 6 shows that it's still a problem.)

ENTITLEMENTS AND RISKS

Consider the legacy credit card system—and in particular, the risks and potential losses to the banks that issue credit cards and the merchants who accept credit cards as payment. Someone who steals a card—or finds a lost card—can make fraudulent charges even without physical possession of the card. An impostor who learns a card's magic numbers and expiration date can make fraudulent charges. A dishonest consumer might make a number of purchases but then disavow them by falsely claiming the card had been lost. A careless consumer might repeatedly lose their card. A financially irresponsible consumer might stop paying the monthly bills.

Despite all this exposure, two key features have enabled this system to persist. First, financial mechanisms exist to move the risk around. Banks can shift the cost of bad transactions to merchants; banks can charge riskier customers higher interest rates and fees; merchants can require minimum charges for transactions. Second, parties can decline to participate once the risks are too high. Banks can refuse credit cards to consumers deemed too likely to default or to lose their cards; merchants can refuse to accept credit cards; merchants and banks can deny specific transactions if conditions are not satisfactory.

But as some of my egovernment clients decades ago were well aware of, these mechanisms do not apply when the service in question shifts to a legal entitlement or other kind of basic right. The security and economic features that make credit cards work did not immediately extend to benefits programs like food stamps—credit card issuers are free to deny the cards to segments of the population for whom deployment is judged to be too risky, but if citizens are entitled to something as a legal right, one cannot deny them just because it's inconvenient.

IN THE SMART CITY

At a NIST workshop on smart cities, speakers posited visions of roads and bridges wirelessly identifying passing vehicles and charging tolls to their owners' credit cards. Will the IoT close off public infrastructure to those without credit cards?

If standard treatment for healthcare evolves to use IoT-based monitoring at home, will citizens be denied treatment if they live in an area with poor connectivity—or themselves cannot afford home WiFi?

If sophisticated cryptographic aggregation and blinding requires sophisticated computing technology in one's smart grid home, will only the affluent have privacy?

Klint Finley in *Wired* wrote how this smart future may "leave many people behind" [5]:

> *Developing nations—precisely the ones that could most benefit from IoT's environmental benefits—will be least able to afford them.... [T]he IoT could lead to a much larger digital divide, one in which those who cannot or choose not to participate are shut out entirely from many daily activities. What happens when you need a particular device to pay for items at your local convenience store?*

The IoT Enforcing Preexisting Socioeconomic Divides

The preceding sections considered various ways the IoC and IoT can lead to divisions between the connected class and the disconnected class—divisions that may perhaps somewhat align with preexisting socioeconomic divisions.

However, for some applications, the division may be more explicit, almost seeming to result from a conscious choice to optimize for one group of people over another. I've talked to seniors who feel that way already about being unable to get traditional printed media from public libraries, or who have to go online to get tax manuals. Over the decades, I've regularly encountered services (such as medical questionnaires from my primary care physician or grant paperwork from universities) that implicitly assume the only possible machine one can use is a Windows PC and the only possible browser is Internet Explorer. In April 2016, Gerry McGovern wrote in *CMSWire* of the consequences of optimizing for electronic in retail [8]:

> *Digital self-service is a double-edged sword. Although it reduces costs, it creates distance between customer and organization. From a customer perspective, self-service means behaving in a semi-automatic, instinctive way.... Designing for this sort of environment requires an incredibly deep understanding of human behavior. Yet, as organizations roll out self-service they get rid of the very employees who actually understand and regularly deal with customers.*

I remember in the early days of ATMs in the US, banks would charge customers a fee to use the ATM. In recent years, banks have instead started charging customers to use tellers. Will the grocery store or bookstore or post office start charging if I don't use the electronic version?

Will smart infrastructure serve to ease life for the affluent at the expense of the poor? For example, consider the Surtrac smart traffic light system developed at Carnegie Mellon in Pittsburgh, Pennsylvania (e.g., [10]). Pollution and wasted time are significant costs of automobile traffic, and using IT to coordinate traffic lights to shape traffic flow to reduce these costs sounds like a good thing. The research presentations discussed impressive results at pilot sites—but then I noticed where the pilot sites were (Penn Circle and East Liberty, in Pittsburgh). When I lived in Pittsburgh as a graduate student a few decades ago, East Liberty was a neighborhood in which one was very careful at night, and Penn Avenue/Route 8 was the corridor connecting the university and medical neighborhoods of the city through disadvantaged and dangerous Wilkinsburg to nicer suburbs. The pilot sites were probably chosen because those immediate areas have gone through some recent redevelopment, and so provided a nice opportunity to insert prototypes into real infrastructure. On the other hand, I saw the numbers about improved wait times and wondered: whose lives were being optimized? Potentially, such systems could make life wonderful for the affluent commuters coming through disadvantaged neighborhoods, at the expense of the people who live there.

Another area that may require design choice is smart medicine. IoC (and eventually IoT) medical applications can potentially improve healthcare and make it accessible to wider populations. However, health informatics researchers such as Kay Connelly at Indiana University point out that many of the populations these services try to reach are educationally disadvantaged or even "functionally illiterate"—and designing effective web and mobile interfaces for such groups is substantially different from designing them for other demographics (e.g., [2]). Even the assumption that a personal cellphone is indeed a personal avenue of communication comes into question.

Another IoT-style domain where design requires a demographic choice is EMRs for children's hospitals. As my colleague Ross Koppel of UPenn has documented, data details for clinicians treating children can have significant and safety-relevant differences from data about general patients. For one example, age may need to be expressed in hours or even minutes—and perhaps even as a negative, for patients still in utero. For another, medication may critically depend on body weight, so the body weight units need to be clear (kilograms or pounds?), and dosages need to be clearly indicated as "mg per kg" or "mg total." However, clinicians lament that children's hospitals are an "orphan subset" of the EMR market in the US—and it's not economically worth it for a vendor to design specifically for that demographic.

The IoT Creating Divides Among Connected Classes

Another avenue to consider is the role of the IoT in promoting—or splintering—social cohesion.

Even in the early days of "blitzmail" and electronic life at Dartmouth, it wasn't clear whether this new thing helped or hurt the sense of community. On the one hand, students might bury their noses in computer screens instead of actually meeting and doing things with other students. On the other hand, students might connect with other students with shared interests but who otherwise moved in different social circles.

As the IoC progressed, the same issues continued. Does "social" networking actually promote or hinder social connection? The common meme and cartoon motif of people together ignoring each other but paying attention to their cellphones suggests one answer. An emeritus professor of social science once quipped to me about the conflict between IT and human evolution: to paraphrase, "How can you work on a team with people when you don't even know how they smell?" Is ecommerce destroying or rescuing Main Street? The conventional wisdom is the former—why pay higher prices at a local shop when Amazon Prime can bring it to you cheaply in two days? On other hand, some analysts argue that by opening up market connections to the whole world, ecommerce can help a specialty business in a small town stay alive.

As we move into the IoT, what else will happen?

One area of concern is the consequences of the transformative power of the IoT on media. Will the end-user convenience (and advertiser delight) of individually customized radio stations and newspapers and billboards and even logos on sports uniforms shatter the sense of community, previously held together by common media experiences? Media can define and shape community. Go back a few decades, and cities only offered a few newspapers and a few mainstream TV and radio stations. The fact that we were all reading the same newspaper, listening to the same drive-time DJs, watching the same nightly newscast all created and contributed to a sense of "us": the local community sharing this experience. Advertising and wider-scale broadcasting still helped with this: we remember a particular outrageous clothing store ad "we all saw"; the local businesses advertised on the rink sideboards on the televised NHL game gave a sense of where that place was.

As the IoT transforms media by enabling fine-tuning of content and advertising to the end consumer, society risks losing these connections. Rather than sharing the same drive-time DJ, we each listen to our own personal Pandora stations. Rather than sharing in a common presentation of the news, we read or watch items handpicked for our own interests (and political beliefs). We may even see different advertisements on the rink boards and outfield walls; our own personalized ghost advertisements and businesses appearing as our smart glasses sense our location or context. What will happen to the sense of community? Will we be citizens of Town X, or of Internet Chat Channel Y? Will we care as much about the welfare of our neighbors if we have fewer connections to them?

(On the other hand, as with Main Street, this connectivity may also serve to promote geographically diverse communities. Besides computer-generated streaming, actual human-hosted radio shows focusing on specialized genres and specialized podcasts stream on the internet. The IoC makes it possible for more people to form a personal weekly connection with WGBH's Brian O'Donovan.)

As McGovern observed [8]:

> *As people grow closer to their friends, family and peers through use of digital, they grow more distant from the establishment, brands and organizations.*

Self-driving cars may have a similar effect. Instead of connecting to the streets through which they drive ("Hey, maybe I should stop at that coffee house") or feeling part of the community of other commuters, drivers will instead be absorbed in their own virtual worlds.

Looking Forward

As we move into the IoT, how can we mitigate these social divisions?

Early in the chapter, we discussed the divisions following from the basic lack of network plumbing. One way to reduce these divisions is to promote more plumbing, just as in previous generations societies promoted telephony and electricity and running water. Many private and public advocacy groups are doing just that. Another angle would be to jump to a new technology (8G?) that will eliminate the obstacles faced by lack of physical wire.

Lack of technological infrastructure in a society also creates an opportunity—because with it comes a lack of inertial resistance from the established infrastructure. Arguably, this may be why cellphone technology caught on first in the developing world—and why some predict that innovative smart grid technology may first catch on in developing countries that do not have a large investment already in the traditional grid. Maybe the IoT can overcome digital divides via such leapfrogging.

The drive for profits may sometimes cloud technological development, but it can also drive players to look for strategies to reach across digital divides (in other words, to reach new markets). The Huawei report stressed the need to reach these groups [7]:

> *Business models that create value are critical, even if what's on offer is "free." Poor and disadvantaged groups often targeted for digital enablement should be treated like any customer. They need to be convinced that they can benefit in order to "invest" in a digital enablement solution, whether it actually costs them money or not.*

Huawei and others also stress the role of increased "digital literacy" to bridge the divide.

If participation in the IoT becomes an implicit part of the human experience, we need to make sure that everyone has the option to be fully human.

Works Cited

1. J. Anderson and L. Rainie, *The Internet of Things Will Thrive by 2025*. Pew Research Center, May 14, 2014.
2. B. Chaudry and others, "Mobile interface design for low-literacy populations," in *Proceedings of the ACM SIGHIT International Health Informatics Symposium,* January 2012.
3. S. P. Crawford, "The new digital divide," *The New York Times,* December 3, 2011.
4. C. Dade, "Why new photo ID laws mean some won't vote," *National Public Radio,* January 28, 2012.
5. K. Fineley, "Why tech's best minds are very worried about the Internet of Things," *Wired,* May 19, 2014.
6. J. Goodman, "The digital divide is still leaving Americans behind," *Mashable,* August 18, 2013.
7. Huawei, *Digital Enablement: Bridging the Digital Divide to Connect People and Society*. 2015.

8. G. McGovern, "The new digital divide," *CMSWire*, April 11, 2016.

9. M. C. O'Connor, "IoT on the farm: Bridging the digital divide," *IoT Journal*, May 19, 2015.

10. S. F. Smith and others, *Real-Time Adaptive Traffic Signal Control for Urban Road Networks: The East Liberty Pilot Test.* Carnegie Mellon University Robotics Institute Technical Report CMU-RI-TR-12-20, 2012.

11. S. W. Smith and R. Koppel, "Healthcare information technology's relativity problems: A typology of how patients' physical reality, clinicians' mental models, and healthcare information technology differ," *Journal of the American Medical Informatics Association*, June 2013.

12. R. H. Weber and R. Weber, *Internet of Things Legal Perspectives.* Springer, 2010.

13. The White House, *Mapping the Digital Divide.* Council of Economic Advisors Issue Brief, July 2015.

The Future of Humans and Machines

The IoT deepens and changes the interconnections between human space and cyberspace. To conclude, this chapter considers some of the deeper implications of these changes by presenting a semiotic framework for this interconnection, using this framework to examine the IoC to IoT transition, then examining other issues—ethics, boundaries, economics, life—relevant to being human in the new information age.

A Framework for Interconnection

In the coming IoT age, newfangled computer systems will smash into human processes. However, as Chapter 3 noted, this future has been here before. Much of my research work (and, before I returned to academia, my professional work) focused on the issues that arise when this occurs. What are the security and privacy issues if a citizen-facing government agency offers services over the web? Why does adding EMR systems to a hospital sometimes cause problems instead of solving them? Why do password rules intended to improve security actually make it worse?

SEMIOTIC TRIADS, IN 2013

This work led to a framework. When my colleague Ross Koppel and I were analyzing usability trouble in health IT [27], we found it useful to consider three things:

- The mental model of the clinician working with the patient and the health IT system

- The representation of medical reality in the health IT system
- The actual medical reality of patients

Figure 10-1 illustrates. (This model could clearly extend to include other actors with mental states—for example, the patients.)

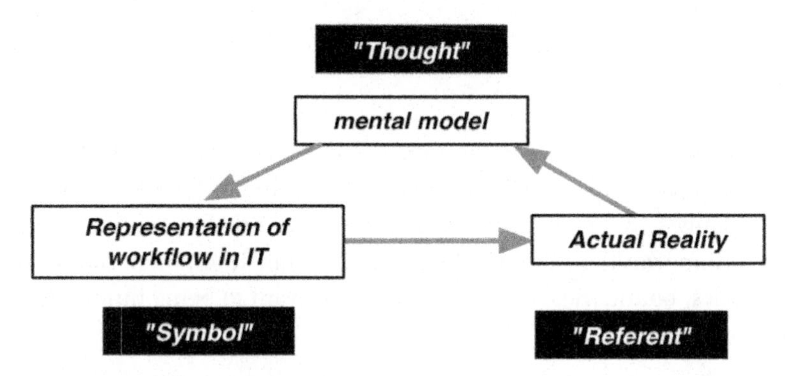

Figure 10-1. The basic Ogden–Richards triad, moved into 21st-century IT; the arrows indicate the main direction of mappings. (Adapted from my paper [26].)

In theory, these three things should show an exact correspondence. The IT system should actually describe the medical reality; the output of the IT system should enable the clinician to infer correct facts about medical reality; inputs to the IT system should correctly reflect clinician actions affecting the reality of the patient. With advances in medical IT, the IT may even directly interact with reality: computers may control the drug dosage administered by a "smart pump"; sensors on a patient may input data directly into the patient's electronic record.

In practice, we found that in the usability problems we identified in our fieldwork, there was a lack of correspondence. Usability problems organized nicely according to mismatches between the expressiveness of the representation "language" and the details of reality—between how a clinician's mental model works with the representations and reality.

SEMIOTIC TRIADS, IN THE 1920S

Somewhat to our chagrin, we discovered we had been scooped by almost a century. In their seminal 1920s work on the meaning of language, Ogden and Richards [22] constructed what is sometimes called the *semiotic triad*. The vertices are the three principal objects:

- What the speaker (or listener/reader) *thinks*
- The *symbol* they use in the language
- The actual item to which they are *referring*

Much of Ogden and Richards's analysis stems from the observation that there is not a direct connection from symbol to referent. Rather, when speaking or writing, the referent maps into the mental model of the speaker and then into the symbol; when reading (or listening), the symbol maps into the reader's (listener's) mental model, which then projects to a referent, but not necessarily the same one. For example, Alice may think of "Mexico" when she writes "this country," but when Bob reads those words, he thinks of "Canada"—and (besides not being Mexico) his imagined Canada may differ substantially from the real one. Thanks to the connection of IT and reality, we now have a direct symbol–referent connection, complicating the merely linguistic world Ogden and Richards explored.

The semiotics of language and the effective communication of meaning focus on *morphisms*—"structure-preserving mappings"—between nodes of the triad. However, with IT usability problems we are concerned instead with ineffective communication and hence focus on what we called *mismorphisms*: mappings that *fail* to preserve important structure when we go from z in one node of the triad to its corresponding z' in another. (See Figure 10-2.) Indeed, we later explored the mismorphisms that lie at the heart of user circumvention of security control, because they characterize the scenarios that frustrate users—and often the resulting circumvention itself [26].

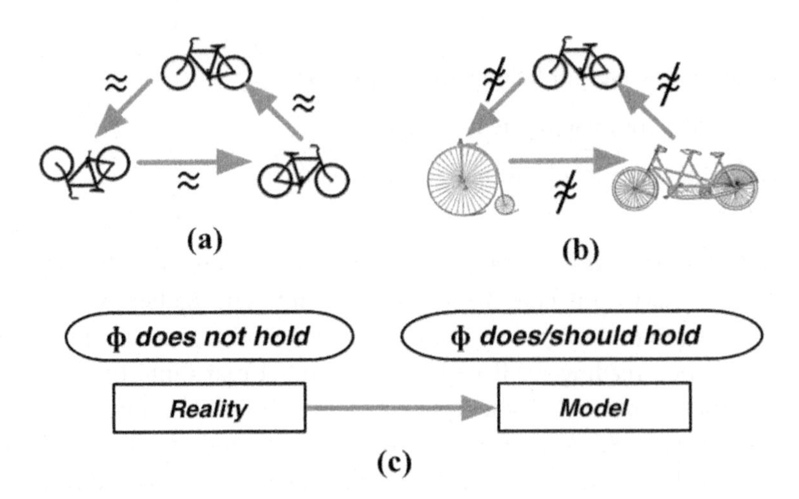

Figure 10-2. (a) Standard semiotics considers structure-preserving mappings between the nodes of the triad. (b) In circumvention semiotics, we think about mappings that fail to preserve structure. (c) For example, in a standard mismorphism scenario, the generated reality fails to embody a property the user regards as critical. (Adapted from my paper [26].)

To reestablish context: when we have computers in real-world applications, we can see the computation as a depiction—in signs and symbols—of reality. The computing reflects assumptions and beliefs about how things work. Consequently, when we have computers and reality and humans involved, we have the opportunity for semiotic confusion. The developers produce systems and code (signs and symbols) that reflect their own mental model of what the real world is; these are different things.

Human/Machine Interconnection in the IoT

The emerging IoT runs the risk of taking this confusion to a whole new level, for several reasons:

- The IoT will tie orders of magnitude more computers to vastly more parts of physical reality.
- Unlike the linguistic scenarios of the last century, the IoT will greatly expand the direct connection between computing and reality (even without human intermediaries).

- The vast decentralization and potential permanence of IoT devices leads to vastly more chances for texts to be wrong—and for wrong texts to remain uncorrected.

As a result, this triad becomes a convenient framework for looking at how these new machines will affect humanity.

MAPPING, LITERALLY

The preceding discussion talked about "mapping" in the mathematical sense (as a "function" or "correspondence"). A computer in some real-world application (say, monitoring the health of a human) has internal state; the configuration of this internal state should (in theory) correspond to the configuration of the human.

However, many IoT applications require literal mapping: representation of the actual real world. For example, Google's extensive mapping database is a key part of the secret sauce that makes Google's self-driving cars possible. In space alone, we already see interesting incidents of friction between things as they are and things as the computers think they are.

The region of Vermont and western New Hampshire where I live regularly provides amusing and mostly harmless examples. Over the last two centuries, the rural areas here underwent "negative development," as people discovered farming was easier in the flatlands of the Midwest and moved away. Farms, houses, and towns were abandoned; forests grew back. However, the *roads* have persisted: sometimes as legal rights of way, sometimes not; sometimes on maps, sometimes not; sometimes traversable with an ordinary car, sometimes only with a four-wheel-drive car, or only with a full-suspension mountain bike, or only on foot. These "ancient roads" lead to lots of fun. Google Maps used to direct some travellers to drive on one that goes underneath the Hanover Reservoir, and to send others on ones that go through an abandoned copper mine. In summer 2016, it gave me directions to an Appalachian Trail parking area by having me drive there on the Appalachian Trail (neither legal nor possible); today, it still shows Goss Road going over Moose Mountain, even though dedicated hikers have not been able to find any evidence of the route on the ground.

A few times a year, local news reports on tourists who need to be rescued because they followed their fancy GPS navigation devices and drove their cars places where cars really shouldn't go. On a more serious note, international news also reports on drivers who have died this way: rather than merely being stuck on rocks or snow, they drove cars into water, or into remote backcountry where rescue was not possible.

Human migration isn't the only cause of mapping error here. In summer 2016, the Australian Broadcasting Corporation reported how Australia's GPS coordinates are wrong by "more than 1.5 meters," and noted that Geoscience Australia's Dan Jaksa observes "with the applications that are coming in intelligent transport systems—like driverless cars—if you're 1.5m out then you're in another lane" [4]. Also in summer 2016, Niley Patel wrote how even more rudimentary mapping problems hamper smart infrastructure such as self-driving cars [23], due to "the 'egress problem'—the way we locate buildings on a map doesn't really describe how people move in and out of those buildings."

MAPPING, FIGURATIVELY

Embedding IT in the real world raises another mapping challenge: encoding real-world processes and flow in the IT algorithms and architecture. Computers only do what they are told to. But as anyone who has ever been frustrated by inept bureaucracy knows, simply following rules is extremely problematic when the process they represent isn't so simple.

The IoC has already shown this frustrating friction.

One famous area is digital rights management (e.g., [2]). Until the latter part of the 20th century, artistic creations (such as texts) were explicitly grounded in the physical media (such as paper) underlying them. Humanity had millennia to evolve understandings of reasonable behavior with regard to such art and the underlying media and to see that the two coincided, more or less. However, the explosion of digital media changed all that, as stakeholders such as the recording industry, YouTube, and scholarly reviewers seeing massive copy-and-paste plagiarism can attest. DRM emerged as an attempt to have technology reintroduce "correct" behavior to this new digital media. However, among computer professionals, DRM is widely regarded as simply not working (e.g., [11]). We regularly see sad but amusing incidents where a website takes down material due to a robot-generated copyright infringement claim that is incorrect but automatically believed. Codifying all the nuances about fair use and such is surprisingly hard to do—particularly a priori. Quoting someone else, Ed Felten famously lamented that "computers are too stupid to look the other way" [13]. (This isn't to say that the problem of knowing when to look the other way is not something that *could* be solved by a computer—indeed, colleagues who assert that the human mind is nothing more than a computer made of meat would argue that we already have working examples! I'm not sure if I would go that far—but I am comfortable asserting that the problem of codifying such behavior in a way that our human-made computers can carry it out has turned out to be far, far more complicated than envisioned.)

For that matter, the onslaught of lawsuits about online music piracy themselves (e.g., [1]) rest on a mapping problem: IP addresses do not equal human actors.

Another area that's less famous (although I'm trying to correct that) is that of circumvention of computer controls and processes. Over and over again, we see scenarios in IT-enhanced workplaces where ordinary users, trying to get their jobs done, circumvent the security controls in the IT. For just a few examples from my previous papers [26, 27]:

- *Some smart pumps assume the patient never weighs more than 350 pounds; for overweight patients, clinicians must similarly distort the IT representation by hooking up two pumps (each allegedly serving a patient of half the actual weight), or by telling the pump it is giving a different medication that, for a legal weight patient, works out the correct drip rate.*

- *A vendor of power grid equipment had a marketing slide showing their default password and the default passwords of all the competitors. The slide was intended to show how secure this vendor was, since they used a more secure default password. However, a deeper issue here is that access to equipment during an emergency is critical, since availability of the grid is far more important than other classical security aspects. Any scheme to replace default passwords with a stronger scheme needs to preserve this availability.*

- *For some medications, a clinician may need to prescribe a tapered decline (sometimes called staged reduction) of dosage rather than an abrupt end. However, the EMR IT does not allow for a taper; what the clinician thinks of as a single unit—the tapered end of medicine—must be instantiated as a sequence of separate non-tapered medication orders, with the clinician needing to remember to terminate the earlier items in the sequence.*

One medical clinician told of screen-scraping a medical record into PowerPoint and then emailing the result to a colleague for a second opinion, because her hospital's data access policy did not allow what her ethical duty required. Another medical clinician even asked us [24]:

> *Are you trying to build a better policeman, or do you want to help patients? Because they're not the same thing!*

(Those trained in the traditional "confidentiality/integrity/availability" definition of security would assert that forgetting to provide the core purpose of the system violates the third principle of security: availability.)

Even in these current IoC (and IoT) application domains, trouble arises because the workflow embedded in the IT does not match the workflow in the real world. In the future IoT, when the tie between IT and the real world is even more intimate and ubiquitous, what will we see? Maybe even more trouble—or maybe we (the computer science community) will learn from our past mistakes. My own team's circumvention work suggests two potential promising avenues:

- *Closing the loop.* Don't just stop with the IT codification of the real-world process: measure if it actually works. (When we asked a senior clinician at a major New York hospital whether the IT developers had any idea of the trouble their incorrect assumptions had caused, he said: "Not at all.")
- *Allowing for override.* When end users perceive that the system does not match reality, let the end users change the system. (At least this way, the system actually knows what's happening; with circumvention, the system departs further from reality.)

UNCANNY DESCENTS

Mismorphism can hamper reasoning about IoT applications in deeper ways, as well. Many application areas might have some desirable and measurable property and some kind of tunable parameter that, in theory, affects this property. For example, my earlier work on security circumvention considered password-based authentication; here, the property might be something such as "net aggregate security," and the parameters might be such things as "minimum password length" or "frequency of required password changes." In a human's mental model, turning these parameters "up" should make the property go up. However, in reality, dialing up the parameter can make things worse. Mapping from IT to reality loses a fundamental property.

My team started using the term *uncanny descent* for this kind of mismorphism, inspired by computer graphics' use of the term *uncanny valley* for when dialing up realism makes things worse before it makes things better. But since we don't know whether things will get better, we stick with just one slope.

Many IoT applications already in deployment demonstrate uncanny descents.

Aging in place

One family of "smart home" applications receiving much research attention is *aging in place*. Helping older people stay in their own homes longer (rather than moving into retirement and assisted-care facilities) can both improve their quality of life and save them money. However, staying in one's own home can increase risks—what if the elder suffers a sudden health-related crisis? To mitigate these risks, researchers have been exploring the use of IT; for instance, augmenting the elder's household with telecommunications devices that let remote relatives and caregivers monitor their health.

The idea here is that adding IT to the household will help improve quality of life. However, researchers at Indiana University observe that it may have the opposite effect [16]:

> *Finally, such systems may have the unintended consequence of reducing the number of phone calls or visits from caregivers, because the caregiver now knows that the older adult being monitored in their home is safe and secure for the present moment.... A primary concern that older adults express about these types of in-home technologies is that they will replace human contact with formal and informal caregivers.*

The conclusion here is not to give up, but rather to be aware—to keep in mind the overall goal of the smartening, and to measure and tune appropriately.

Self-checkout

Another area receiving attention (and hype) is incorporation of smart technology into retail shopping. For myself (and probably many readers), a tangible manifestation of this trend is the recent proliferation of self-checkout stations at grocery stores. The idea here is that adding IT to this part of retail can improve revenue due to quicker checkout and reduced costs (since fewer employees are needed).

However, again, researchers Adrian Beck and Matt Hopkins at the University of Leicester have noted it may have the opposite effect [20]:

> *One million shopping trips were audited in detail, amounting to six million items checked. Nearly 850,000 were found not to have been scanned, the report said, making up 4 percent of the total value of the purchases.*

Whether the items were unscanned due to intentional theft or inadvertent error is not clear.

Self-checkout is also a wonderfully tangible example of a mismatch between the real-world process expected by a shopper and the way this process is codified inside the smart checkout system—with (as Beck observed) "the phrase 'unexpected item in the bagging area' striking dread into many a shopper."

In the IoT as with energy, friction generates waste.

Safety

Adding smart technology to automobiles is intended to make them safer, but may have the opposite effect. *Freakonomics*'s Steven Dubner quotes Glenn Beck [12]:

> *I was looking at an Audi as well, and the guy said to me, he said, "this has some amazing safety features, it knows when the car is going to roll, if your window is rolled down, it immediately rolls the window up, it has the side airbags, your seats, depending on what the car senses it's going to do, it puts the seats put in right position," you know, it makes me want to flip the car! I'm going to put my seat in the most awkward position, and I'm gonna flip it! This is, like, the safest car on the road, he used the term "death-proof." But honestly I didn't even think about it until we were—until I was driving it. And I thought—I really was taking a corner a little too fast, and I'm like "I can handle it, what's the worst that can happen?"..."What? So I didn't stop at the stop light, and I'm going a hundred and ninety? What? I can flip it, I'll survive, it's the death-proof car!" What a dope!*

As a middle ground between traditional automobiles and self-driving automobiles, Liviu Iftode and his colleagues have been considering the challenges of *remote driving*: that is, a car on the road that is in fact "driven" by a human at a remote control center (e.g., [19]). Besides the engineering challenges, this concept also raises psychological ones, as the physical separation isolates the human pilot from the physical consequences of their actions. Will separation reduce the propensity to road rage and cause the pilot to drive more safely? Or, as with Glenn Beck's death-proof Mercedes, will separation reduce the sense of personal risk and cause the pilot to drive more dangerously? (Similar questions have been raised about the use of remotely controlled drones in combat.)

Of course, this problem predates smart IT. Studies of antilock brakes in automobiles suggest that this safety-promoting technology can increase the risk of accidents because it also promotes more "aggressive" driving [28]. In American football, the use of pads and helmets can similarly increase the risk of injuries, both because they decrease the perceived risk of injury to a player and because (with helmets, at least) the safety devices themselves can be used as weapons. As Dubner observes: "As the safety equipment gets better, our behavior becomes more aggressive" [12].

In the workplace

Uncanny descents have been seen when moving IT into the workplace too. As a security researcher, I have heard multiple colleagues (all from the public sector) rant about how important it was to block employees from doing personal web browsing during work hours—even though studies have shown that such activity can *boost* productivity [3], essentially since it provides micro-breaks that are more time-efficient than walking to the water cooler. On the other hand, a recent university study "showed that employees' performance improved 26 per cent when their smartphones were taken away" [25].

Others

Some recent items in the news demonstrate scenarios where stakeholders are aware of what they perceive as negative effects of bringing IT into real life—and are taking action.

To further its mission of promoting the right to bear firearms in the US, the National Rifle Association (NRA) has ensured that the US facility where law enforcement traces guns is not allowed to have computers [18]:

> *That's been a federal law, thanks to the NRA, since 1986: No searchable database of America's gun owners. So people here have to use paper, sort through enormous stacks of forms and record books that gun stores are required to keep and to eventually turn over to the feds when requested. It's kind of like a library in the old days—but without the card catalog. They can use pictures of paper, like microfilm (they recently got the go-ahead to convert the microfilm to PDFs), as long as the pictures of paper are not*

> *searchable. You have to flip through and read. No searching by gun owner. No searching by name.*

Cyclists Nairo Quintana and Alejandro Valverde, while sitting in first and second place in the Vuelta a España, called for the banning of power meters (IT in the bicycle, to help measure and then tune rider performance) [14]:

> *"They take away a lot of spectacle and make you race more cautiously," Quintana said. "I'd be the first in line to say they should be banned." "I think they're really useful for training, but they take out a lot of drama from the sport," added Valverde. "In competition you should be racing on feelings."*

Ethical Choices in the IoT Age

In his novel *A Clockwork Orange*, Anthony Burgess considered whether moral decisions made by clockwork would have the same value as moral decisions made by unconstrained humans.[1]

In an IoT world, many scenarios exist where choice and action move from a human in a direct situation to a machine, or perhaps to the human who programmed that machine at some point long ago. Some of these choices will inevitably have moral and ethical dimensions. How will the reality—and the human perception—of the choices change when the actors move?

Ethicists have long considered the *Trolley Problem* (e.g., see Figure 1 in [15]). One can construct many scenarios involving unfortunate arrangements of trains and human bodies in which a human actor must make a choice between two options. For example:

- If a runaway train is about to hit a fork where it can go in one direction and kill five people or can go in another and kill one, and Alice controls the switch at this fork, which option should she choose?
- If a runaway train is about to hit five people, but Alice is standing on a bridge over the tracks next to a very large Bob, should she let the five peo-

1 Of course, this was back in an era when most intellectuals believed in free will; such belief seems unfashionable in many modern scientific circles. I think it's safe to say this issue may be beyond the scope of this book.

ple die, or push Bob off the bridge, so he dies but stops the train and thus saves the other five?

Psychologists have studied why humans tend to make different choices in the different scenarios, even though (when evaluated solely in terms of net utility, or lives lost) the choice in each scenario should be the same—lost one life to save five. There's a mismorphism here: something changes when the situation is mapped from end reality to human perception of the action.

Now, fast-forward to a future of smart technology, when the actor is no longer human. If a self-driving car is in an unfortunate scenario where it must kill either five pedestrians in one lane or one in another, which should it choose? What if the choice is between killing five pedestrians, or crashing into a wall and killing its driver? And what about all the other kinds of smart machine actors the IoT will bring?

These questions are grist for much philosophical discussion (e.g., see [10]).

Perception of Boundaries in the IoT Age

Another structure that can be lost in the mapping between smart IT, the real world, and human perception is that of boundaries.

Even in the plain IoC, the borders in IT infrastructure do not match the borders in reality. Back at the turn of the century, Bill Cheswick's Internet Mapping Project (*http://cheswick.com/ches/map/*) revealed internet connectivity suggesting mergers between companies that had not yet announced they were merging, and connectivity suggestive of covert nation-state influence. In modern times, Doug Madory of Dyn (*http://research.dyn.com/author/dmadory/*) has done analysis of internet routing data indicating interesting connections with modern political events, among other things; Dyn even markets "internet intelligence" as a business.

As we move into the IoT, as Chapter 1 noted, humans may have difficulty perceiving the risks that compromise of a particular IoT infrastructure, such as smart meters, can pose to apparently unrelated systems, such as the cellphone network; likewise, human perception of security risks to a particular IoT infrastructure, such as Target's point-of-sale terminals, may not take into account attacks launched at apparently unrelated systems, such as HVAC controls. Chapter 4 noted an instance where a port for a car's CAN bus—through which one can unlock the vehicle—can be found outside the locked perimeter. In summer 2016, a security analysis of the massive privacy breach at Banner Health noted "it was odd that the point of sale systems at Banner's 27 food service locations that were affected appear to have been on the same network as clinical systems" [8].

The business relationships in IT manufacturing—who buys and repackages what software and components from whom—also complicate accurate perception of and reasoning about boundaries. As Chapter 4 noted, vulnerabilities in the firmware of one CCTV device affect dozens of vendors, who simply repackaged that device. The ThinkPwn exploit for a low-level UEFI driver from one particular small vendor affects the larger machines (such as some Lenovo laptops) that used that driver [9]. The flaws enabling hackers to shut down Andy Greenberg's Jeep were not in the Jeep itself but in a radio manufactured by someone else—and used in other brands of cars as well [21].

Human Work in the IoT Age

Humans often define themselves by work—"your work is your worth." The IoT changes the workplace. Does it change human worth?

As noted previously, mixing technology into the workplace changes the workplace. One might think that it would make things "better," by some definition: people could get more things done more accurately in less time. For example, NPR's *Planet Money* notes [17]:

> *The economist John Maynard Keynes predicted [in 1930] that his grandkids would work just 15 hours a week. He imagined by now, we would basically work Monday and Tuesday, and then have a five-day weekend.*

Why did Keynes get this wrong? NPR posits it was because although technology improved productivity, people still choose to work, due to some combination of opportunity cost (that hour of leisure is not worth the lost income) and personal satisfaction. Observers taking a long view of the US economy might add that even the productivity improvement cannot compensate for the decrease in wages per unit of productivity, and the increase in cost of living. Back in the 1980s, the older businessman who ran one of the first tech companies I worked for lamented how (even then) both parents working would not support a family as well as one parent working when he was younger. Paraphrasing, "I never thought I would see the next generation do worse than the previous one."

The IoT promises an even more disruptive technological revolution in the workplace. In terms of the triad discussed earlier, the permeation of IT into physical reality fundamentally changes the way business processes work—and human understanding of these processes (and of the role of human workers in it) lags behind.

In particular, the new technology enables a sort of arbitrage: tasks that required specialized humans can now be done by machines; tasks that required expensive local infrastructure (such as big computing) can now be outsourced inexpensively (e.g., to the cloud). In a series of articles [5, 6, and 7], writers Bernard Condon, Jonathan Fahey, and Paul Wiseman analyzed this impact:

> *Five years after the start of the Great Recession, the toll is terrifyingly clear: Millions of middle-class jobs have been lost in developed countries the world over. And the situation is even worse than it appears. Most of the jobs will never return, and millions more are likely to vanish as well, say experts who study the labor market. What's more, these jobs aren't just being lost to China and other developing countries, and they aren't just factory work. Increasingly, jobs are disappearing in the service sector, home to two-thirds of all workers....*
>
> *For more than three decades, technology has reduced the number of jobs in manufacturing....*
>
> *Start-ups account for much of the job growth in developed economies, but software is allowing entrepreneurs to launch businesses with a third fewer employees than in the 1990s....*
>
> *Those jobs are being replaced in many cases by machines and software that can do the same work better and cheaper....*
>
> *Reduced aid from Indiana's state government and other budget problems forced the Gary, Ind., public school system last year to cut its annual transportation budget in half, to $5 million. The school district responded by using sophisticated software to draw up new, more efficient bus routes. And it cut 80 of 160 drivers....*

The analysis concludes with considering what will happen in a society where (thanks to technology) a majority of humans cannot find employment—or must compete for a vanishing number of "midskill" jobs. The machines will be wonderful and enable a wonderful life for the innovators. But what about the rest of us?

> *As far back as 1958, American union leader Walter Reuther recalled going through a Ford Motor plant that was already automated. A company manager goaded him: "Aren't you worried about how you are going to collect union dues from all these machines?" "The thought that occurred to me," Reuther replied, "was how are you going to sell cars to these machines?"*

Many in the field already talk about *dark factories* (with no human employees). However, humans like doing—that's one of the reasons Keynes was wrong. What will our future hold?

Brave New Internet, with Brave New People in It

The previous chapter used the fashionable term "digital divide." A similarly fashionable term, "digital natives" refers to humans who grow up with new information technology. Rather than having to adapt to a new world (as "digital immigrants" must do), a digital native has always seen the universe as having this IT enhancement.

Confronted with the internet, the web, digital music, YouTube, iPods, and smartphones, digital immigrants are often frightened by the skill and ease with which digital natives interact with the digital. As Groucho Marx quipped in technologically ancient times:

> *A child of five would understand this. Send someone to fetch a child of five.*

In slightly less ancient times, when I was considering leaving industry for academia, I asked a mentor how you lead students. His response: "You don't lead them—you follow them."

The IT of the 2010s is considerably advanced from the IT of the 1990s or the 1970s, and the teenagers of the 2010s perceive a very different world from their predecessors. However, the IoT of the 2020s promises (or threatens) to be potentially a far greater advancement. What world will the digital natives of the IoT grow up in? Will they regard our current dumb houses and dumb cars and dumb bridges as hopelessly archaic? Alternatively, if we don't adequately resolve the security and privacy risks, will they be able to cope if their world reverts to a cyber Love Canal?

After all, they are us.[1]

Works Cited

1. R. Beckerman, "Large recording companies vs. the defenseless: Some common sense solutions to the challenges of the RIAA litigations," *The Judges Journal, American Bar Association,* July 2008.
2. L. J. Camp, "DRM: Doesn't really mean digital copyright management," in *Proceedings of the 9th ACM Conference on Computer and Communications Security,* 2002.
3. J. Cheng, "Study: Surfing the internet at work boosts productivity," *Ars Technica,* April 2, 2009.
4. E. Clark, "Driverless cars need Australia's latitude and longitude coordinates to be corrected," *Australian Broadcasting Corporation News,* July 28, 2016.
5. B. Condon, J. Fahey, and P. Wiseman, "Practically human: Can smart machines do your job?," *AP: The Big Story,* January 24, 2013.
6. B. Condon and P. Wiseman, "AP IMPACT: Recession, tech kill middle-class jobs," *AP: The Big Story,* January 23, 2013.
7. B. Condon and P. Wiseman, "Will smart machines create a world without work?," *AP: The Big Story,* January 25, 2013.
8. J. Conn, "Banner Health cyberattack impacts 3.7 million people," *Modern Healthcare,* August 3, 2016.
9. L. Constantin, "Firmware exploit can defeat new Windows security features on Lenovo ThinkPads," *PC World,* July 1, 2016.
10. C. Doctorow, "The problem with self-driving cars: Who controls the code?," *The Guardian,* December 23, 2015.
11. C. Doctorow, *DRM: You Have the Right to Know What You're Buying!* Electronic Frontier Foundation, August 5, 2016.

1 Parts of "A Framework for Interconnection" on page 203 are adapted from portions of my paper [26].

12. S. J. Dubner, *The Dangers of Safety Full Transcript.* Freakonomics Radio, August 13, 2015.

13. E. Felten, "Too stupid to look the other way," *Freedom to Tinker*, October 29, 2002.

14. A. Fotheringham, "Quintana calls for power meters to be banned from racing," *Cyclingnews*, August 30, 2016.

15. M. Hauser and others, "A dissociation between moral judgments and justifications," *Mind & Language*, February 2007.

16. L. Huber and others, "How in-home technologies mediate caregiving relationships in later life," *International Journal of Human–Computer Interaction*, 2013.

17. D. Kestenbaum, "Keynes predicted we would be working 15-hour weeks. Why was he so wrong?," *NPR Planet Money*, August 13, 2015.

18. J. M. Laskas, "Inside the Federal Bureau of Way Too Many Guns," *GQ*, August 30, 2016.

19. R. Liu and others, *Remote Driving: A Ready-to-Go Approach to Driverless Car?* Technical Report DCS-TR-712, Rutgers University, Department of Computer Science, February 2015.

20. C. Mele, "Self-service checkouts can turn customers into shoplifters, study says," *The New York Times*, August 10, 2016.

21. D. Morgan, "Car hacking risk may be broader than Fiat Chrysler: U.S. regulator," *Reuters*, July 31, 2015.

22. C. Ogden and I. Richards, *The Meaning of Meaning.* Harcourt, Brace and Company, 1927.

23. N. Patel, "Self-driving cars aren't going to be so great until we make our maps way better," *The Verge*, August 24, 2016.

24. S. Sinclair and S. W. Smith, "What's Wrong with Access Control in the Real World?," *IEEE Security and Privacy*, July/August 2010.

25. S. Shinde-Nadhe, "Not using smartphones can improve productivity by 26%, says study," *Business Standard*, August 30, 2016.

26. S. W. Smith and others, *Mismorphism: A Semiotic Model of Computer Security Circumvention (Extended Version)*. Dartmouth Computer Science Technical Report TR2015-768, March 2015.

27. S. W. Smith and R. Koppel, "Healthcare information technology's relativity problems: A typology of how patients' physical reality, clinicians' mental models, and healthcare information technology differ," *Journal of the American Medical Informatics Association*, June 2013.

28. C. Winston and others, "An exploration of the offset hypothesis using disaggregate data: The case of airbags and antilock brakes," *Journal of Risk and Uncertainty*, March 2006.

Index

Symbols

A

B

C

D

E

F

G

H

I

Q

R

S

T

U

V

W

X

Z

About the Author

Professor **Sean Smith** has been working in information security—attacks and defenses, for industry and government—since before there was a web. In graduate school, he worked with the US Postal Inspection Service on postal meter fraud; as a post-doc and staff member at Los Alamos National Laboratory, he performed security reviews, designs, analyses, and briefings for a wide variety of public-sector clients; at IBM T.J. Watson Research Center, he designed the security architecture for (and helped code and test) the IBM 4758 secure coprocessor, and then led the formal modeling and verification work that earned it the world's first FIPS 140-1 Level 4 security validation.

In July 2000, Sean left IBM for Dartmouth, since he was convinced that the academic education and research environment is a better venue for changing the world. His current work, as PI of the Dartmouth Trust Lab and director of Dartmouth's Institute for Security, Technology, and Society, investigates how to build trustworthy systems in the real world.

At Dartmouth, many of his courses have been named "favorite classes" by graduating seniors. His book *Trusted Computing Platforms: Design and Applications* (Springer, 2005) provides a deeper presentation of this research journey; his book *The Craft of System Security* (Addison-Wesley, 2007) resulted from the educational journey.

Sean has published over 100 refereed papers; been granted over a dozen patents; and advised over three dozen Ph.D., M.S., and senior honors theses. He and his students have won several "Best Paper" awards.

Sean was educated at Princeton and Carnegie Mellon University, and is a member of Phi Beta Kappa and Sigma Xi.

Colophon

The cover fonts are URW Typewriter and Guardian Sans. The text font is Adobe Minion Pro; the heading font is Adobe Myriad Condensed; and the code font is Dalton Maag's Ubuntu Mono.

CPSIA information can be obtained
at www.ICGtesting.com
Printed in the USA
BVOW06s0842010217
475057BV00013B/74/P

9 781491 963623